D0991523

Liposuction Surgery and Autologous Fat Transplantation

Liposuction Surgery and Autologous Fat Transplantation

Saul Asken, M.D., FAAPRS
Medical Director
Cosmetic Surgery Center of Connecticut
Westport, Connecticut
Assistant Clinical Professor
New York Medical College
Valhalla, New York

APPLETON & LANGE
Norwalk, Connecticut/San Mateo, California

0-8385-5684-1

Notice: Our knowledge in clinical sciences is constantly changing. As new information becomes available, changes in treatment and in the use of drugs become necessary. The author(s) and the publisher of this volume have taken care to make certain that the doses of drugs and schedules of treatment are correct and compatible with the standards generally accepted at the time of publication. The reader is advised to consult carefully the instruction and information material included in the package insert of each drug or therapeutic agent before administration. This advice is especially important when using new or infrequently used drugs.

Copyright © 1988 by Appleton & Lange
A Publishing Division of Prentice Hall

All rights reserved. This book, or any parts thereof, may not be used or reproduced in any manner without written permission. For information, address Appleton & Lange, 25 Van Zant Street, East Norwalk, Connecticut 06855.

88 89 90 91 92 / 10 9 8 7 6 5 4 3 2 1

Prentice-Hall of Australia, Pty. Ltd., Sydney
Prentice-Hall Canada, Inc.
Prentice-Hall Hispanoamericana, S.A., Mexico
Prentice-Hall of India Private Limited, New Delhi
Prentice-Hall International (UK) Limited, London
Prentice-Hall of Japan, Inc., Tokyo
Prentice-Hall of Southeast Asia (Pte.) Ltd., Singapore
Whitehall Books Ltd., Wellington, New Zealand
Editora Prentice-Hall do Brasil Ltda., Rio de Janeiro

Library of Congress Cataloging-in-Publication Data

Asken, Saul.
　　Liposuction surgery and autologous fat transplantation / Saul Asken.
　　　　p.　cm.
　　ISBN 0-8385-5684-1
　　1. Suction lipectomy. 2. Adipose tissues—Transplantation.
　3. Surgery, Plastic. I. Title.
　　[DNLM: 1. Adipose Tissue—surgery—atlases. 2. Adipose Tissue—transplantation—atlases. 3. Suction—atlases. 4. Surgery.
　Plastic—methods—atlases. WO 517 A834L]
　RD119.5.L55A85 1988
　617'.47—dc19
　DNLM/DLC　　　　　　　　　　　　　　87-30828
　for Library of Congress　　　　　　　　　　　CIP

Designer: M. Chandler Martylewski

PRINTED IN THE UNITED STATES OF AMERICA

To all the great surgeons who led me through the path of cosmetic surgery

We restore, repair or conceal those parts of the face which nature has given but fortune has taken away. Not so much that we delight the eye; but that we may buoy up the spirit and restore the mind of the afflicted.

Gaspare Tagliacozzi
University of Bologna, 1597

Contents

Preface

It has been more than a decade since modern liposuction surgery was introduced to the medical community. In that time, great strides have been made in the technique and instrumentation used in its performance.

Education and experience are a *sine qua non* requirement in medicine and particularly in surgery. This textbook-atlas is not intended to cover the entire scope of liposuction, but to present a successful technique developed during the past 6 years of liposuction and fat transplantation performed under local anesthesia in an outpatient surgical facility. As with any procedure, several textbooks should be read and studied by the physician planning to perform liposuction surgery. These should include textbooks or articles not only on liposuction, but also on the anatomy and metabolism of fat, the pharmacology of the drugs used, and the treatment of emergencies that can arise in an operating room.

Any physician intending to perform liposuction should first attend several courses and workshops organized by the American Society of Liposuction Surgery and the American Academy of Cosmetic Surgery. These courses should be followed by observing the procedures on a one-to-one basis with a surgeon well versed in this technique. Only then should liposuction be attempted, preferably on lipomas.

As with any surgical procedure, technical ability is required. Also, liposuction and autologous fat transplantation are truly facial and body sculpting and require a high aesthetic sense. To obtain pleasing results, the physician must be an artist as well as a surgeon.

Acknowledgments

The phenomenal growth of liposuction surgery during the past few years is due to its efficacy. The technique used in performing this highly aesthetic procedure has obviously contributed to its success. No one can claim to be totally original in innovating techniques and I wish to acknowledge my debt to The American Society of Liposuction Surgery with Julius Newman, M.D. its founder and first president and to The Americal Academy of Cosmetic Surgery with Richard C. Webster, M.D., its first president. It is to Dr. Webster that I owe my greatest debt for his teaching and instilling in me his philosophy of cosmetic surgery, conservatism and artful approach.

My sincere thanks to my entire staff without whose help this book would not have been possible and to Mr. Peter Klamkin of Appleton and Lange who offered a great deal of support and showed immense patience during the writing of this text.

History of Liposuction

1

Several years ago, a new method for the removal of unwanted fat deposits made its appearance on the forefront of cosmetic surgery. Introduced in its present state to the cosmetic surgical community by Dr. Yves-Gerard Illouz and further popularized by Dr. Pierre F. Fournier, both of France, liposuction or suction-assisted lipectomy represents the most exciting and revolutionary cosmetic procedure to appear during the last few decades.

Its apparent simplicity has deluded many surgeons into believing that either "anyone can do it" or that it cannot work because it appears so simple. This apparent simplicity is deceptive; facial and body contouring by liposuction demands from the surgeon not only excellent technical ability but a particularly acute artistic sense as well. What started as volume reduction has developed into true facial and body sculpting; to quote Fournier, "lipoextraction has developed into lipoplasty."

In order to understand its present state, a brief history of liposuction is helpful. The first reported use of body contouring by removal of fat through a small incision was in 1929 by the French surgeon Dujarrier, who used a curette to attempt reduction of the calves of a famous dancer. This procedure resulted in injuries to major blood vessels, which later required the amputation of one leg.

The next known attempt to remove fat through small incisions was in 1968 by the American physician Wilkinson practicing in Hawaii.[1] Although he had acceptable results, he abandoned the method because of inadequate instrumentation. Schrudde, a German surgeon, reported his use of an aspiration curette in 1972.[2-5] Because of ensuing complications, his method did not gain the acceptance of the surgical community.

Although most general plastic surgeons attribute the development of the suction cannula to Kesselring, a Swiss plastic surgeon, it was an Italian father-and-son team of surgeons, Arpard and Giorgio Fischer of Rome, who first presented the modern use of a suction device for removal of fat. In 1975 the Fischers devised a suction instrument incorporating a cutting blade. The cannula aspirated the fat through its distal openings and the motorized cutting blade fragmented it, permitting its suction. Later it was realized that the cannula alone, attached to a sufficiently powerful suction unit, could remove the unwanted fat without cutting it. However, the complications that resulted from the initial technique and instrumentation were enough to discredit the procedure. The Fischers presented their innovative technique in the *Bulletin of the International Academy of Cosmetic Surgery* in 1977.[6] One year later, Kesselring and Meyer described modifications of the Fischers' technique and instrumentation using a modified curette attached to a suction device.[7]

Their technique involved undermining a large area, after which the fat was removed by a curetting cannula attached to a suction unit. Because of its inherent dangers, these surgeons restricted this procedure to small localized deposits of fat in the trochanteric area.

The final improvement of the instrumentation is credited to Illouz of France, who made the procedure safe, acceptable, and successful by the use of a blunt cannula. Illouz reported his results in 1978 and 1979 and had his technique first published in 1980.[8,9] The same year, Dr. Norman Martin became the first American physician to study with Illouz and bring his technique to the United States.

It is noteworthy that Illouz first introduced his procedure in the United States at a meeting of the

American Board of Cosmetic Surgery in January 1982. This group was composed of physicians and surgeons of different specialties, including cosmetic, facial plastic, general, dermatologic, and plastic surgery. The first American workshop on liposuction surgery was sponsored by the American Board of Cosmetic Surgery and took place at the Graduate Hospital in Philadelphia, Pennsylvania, in October 1982. In December of the same year, the American Society of Liposuction Surgery was incorporated under its first president, Dr. Julius Newman.

One year later, Fournier discussed the Illouz technique at a meeting of the American Academy of Facial Plastic and Reconstructive Surgery.[10] One of the attending general plastic surgeons, Dr. Gregory Hetter, was so impressed by the technique that he convinced the American Society of Plastic and Reconstructive Surgery to send a blue-ribbon committee to Paris to investigate the value and merits of this "new" procedure. The following is a quotation from their report: "The committee unanimously agrees that suction lipectomy by the Illouz blunt cannula method is a surgical procedure that is effective and safe in *trained* and *experienced* hands and offers benefits which heretofore have been unavailable" (italics added).[11]

Both Illouz and Fournier further improved the design of the instruments, and the procedure soon gained acceptance in the United States. It was popularized by the American Society of Liposuction Surgery, which, with Julius Newman as president, organized the first workshops in the United States for physicians interested in this procedure.

The terminology associated with the technique of suctioning out fat has been confusing and deserves clarification.

Suction-assisted lipectomy, the term used by most general plastic surgeons, is actually inaccurate. It implies cutting out the fat, which is not the case. Some surgeons have gone so far as to copyright the term for commercial purposes.[12,13] Although Schrudde's technique involved a true lipectomy, Illouz's technique does not.

Lipolysis, the term first coined by Illouz, implies lysis of the fat cells before and during the procedure and is not based on clear evidence. It is true that stronger suction will loosen the adipose cells and cause a physical breakdown, with the consequent release of some of the fat cell content. However, most of the adipocytes removed do not undergo lysis. Furthermore, there is no clear evidence that solution injected prior to suction produces lysis of the fat cells.

Liposuction is the term used by most cosmetic surgeons to describe the removal of fat with a cannula, aided by suction.[14] Although etymologically incorrect because of its mixed Greek and Latin roots, it is the most descriptive and most accurate name for the present technique. This term is currently used by most facial plastic surgeons, dermatologic surgeons (in whose domain the skin and its subcutaneous fat clearly falls), and cosmetic surgeons who specialize only in the aesthetic aspects of plastic surgery.

Blunt-suction lipectomy, a term coined by Hetter,[13] represents a contradiction of meanings. *Blunt suction* refers to suction removal of fat using a blunt-tip cannula. *Lipectomy* implies cutting out the fat. This is certainly not the case when one uses a blunt-tip cannula with its side opening properly smoothed and polished. Hetter's term neither aids in understanding the technique nor describes it. His accusation that the motivation to create new terms is "an attempt to self-servingly carve out new 'turf'" is certainly not supported by developing new terminology himself.

Concerning the qualifications of physicians performing liposuction surgery, it must be emphasized that liposuction originated with, and was taught and perfected by, cosmetic, not plastic, surgeons. Neither Illouz nor Fournier, who initially taught the liposuction technique to American plastic surgeons, is recognized as a plastic surgeon by the American Society of Plastic and Reconstructive Surgery. Furthermore, the first article in modern literature on using a cannula to remove unwanted fat was written by a cosmetic, not plastic, surgeon, Dr. Giorgio Fischer. Clearly, no specialty can have a monopoly on a surgical procedure.

Development, Structure, and Metabolism of Fat

2

It is beyond the scope of this book to present a detailed account of the embryology and metabolism of fat, including a study of obesity; the reader is referred to many textbooks and papers written on these subjects. However, discussion of a technique used to remove a specific component of the body, the adipose tissue, would not be complete without a brief survey of the development and metabolism of this tissue.

DEVELOPMENT

The subcutaneous adipose tissue develops from specific reticuloendothelial structures known as primitive organs appearing in the subcutis, during the fourth fetal month.

It was Virchow who first suggested, during the 1850s, that fat cells are a specialized type of connective tissue.[15] This view is still maintained today, with fat being classified as a specialized form of reticular connective tissue. It originates in the subdermal perivascular connective tissue from adipoblasts, cells indistinguishable from the rest, and develops into preadipocytes. The preadipocytes convert into mature fat cells by accumulating triglycerides.[16] Adipocytes grouped in an organized manner form a lobule having its own blood supply with a central artery feeding a capillary network surrounding every fat cell.

These lobules are separated by fibrous septa that, when organized in a large number, form what is known as adipose tissue, present mainly in the subcutis and the perirenal and mesenteric areas. Adipose tissue represents approximately 20 percent of total body weight and because of its multitude of functions some refer to it as an organ, just as skin is considered an organ of the body.[17,18] Contrary to the independent vascularization of the adipose tissue, accumulations of fat cells in other parts and organs of the body are mainly perivascular and are dependent on the vasculature supplying the nearby organ.

Subcutaneous adipose tissue receives its vascularization from the fascial network, which forms a subdermal plexus. From there, branches arise to form a subpapillary plexus, which is the origin of the papillary loops.[19] The fat lobules receive their blood supply from the descending branches of the subdermal plexus. When the adipose tissue is thick (more than 10 mm), it receives its blood supply both from the descending branches of the subdermal plexus, which feed the upper layer of fat, and from the ascending fascial arteries, which feed the lower layer (Fig. 2–1).

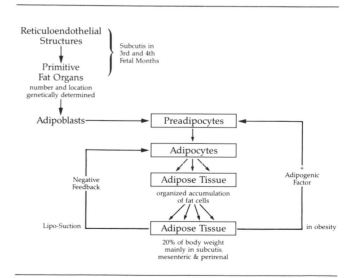

Figure 2–1. The development of adipose tissue from primitive organs is shown.

When this adipose tissue achieves a certain thickness, the descending and the ascending vessels meet at a level where a third subcutaneous vascular plexus is formed. A septum can be formed at this level, thereby forming two relative separate layers of subcutaneous fat.[20] The significance of this anatomic division is unclear, although the gradual increase in size of adipocytes in the deeper layers may be significant for fat transplantation (Fig. 2–2). Further study is required to determine a possible relationship between the size of adipocytes and their survival after transplantation.

The number of fat cells and consequently of lobules increases until physical growth is complete, that is, late puberty to adolescence. This process takes place in a phasic manner, reaching the peak of the curve in late childhood. By contrast, the size of the cells continues to increase to a certain limit throughout life, depending on the diet.[21]

As the location of primitive fat organs is genetically determined and fat is not evenly distributed, the thickness of the subcutaneous fat will vary considerably in different parts of the body. The thickness of adipose tissue varies not only from one part of the body to another, but also from one person to another, giving particular shape to the adult.

The genetic influence on subcutaneous fat has been further confirmed both experimentally and clinically. Transplantation of primitive fat organs results in the formation of adipose tissue in the recipient site; similarly, transplanted flaps retain the donor site characteristics.[22–26]

Gender is another factor that influences the localization of fat. Women have a gynecoid distribution in the lower torso and thighs; men have an android distribution, with the fat deposited particularly in the upper part of the body (Figs. 2–3, 2–4).[27,28]

OBESITY

Obesity has been defined as a 15 to 20 percent increase in weight due to excessive adipose tissue. Overfeeding experimental animals with a high-fat diet during the period corresponding to human childhood, that is, during development, results in a marked increase in the number of adipocytes formed and in the consequent overweight of the animals during adulthood.[29] If this translates to the

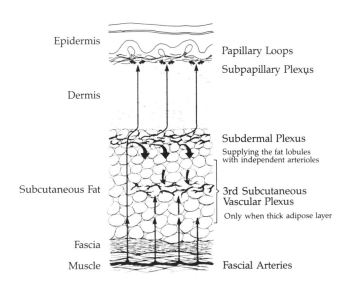

Figure 2–2. This schematic drawing illustrates the vascularization of the subcutaneous fat.

human, a more balanced diet during childhood can have a salutory effect on the health of the adult, particularly in view of the recent evidence of coronary heart disease related to an overweight condition (Fig. 2–5).[30–35]

The study of obesity is further complicated by the fact that android obesity is more common in men and is frequently associated with a greater risk of cardiovascular disease and diabetes. Gynecoid obesity in women does not have the same relationship with either cardiovascular or other metabolic diseases. However, android obesity in women, whether because of its masculinizing origin or not, can be associated with both cardiovascular and metabolic disease.[28]

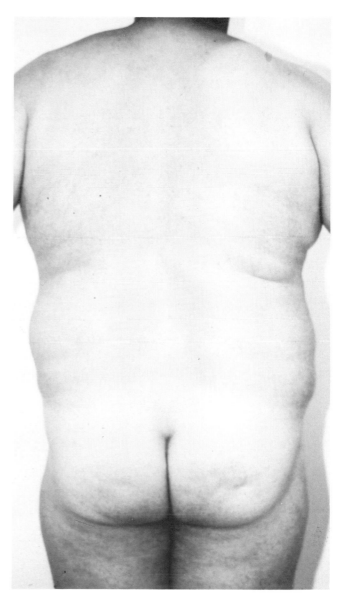

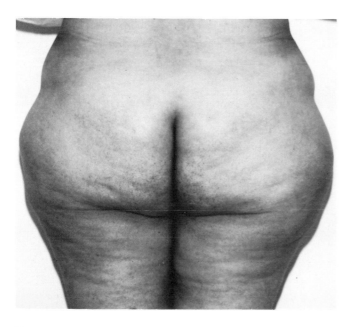

Figure 2–3. Gynecoid (female) obesity is illustrated.

Figure 2–4. Android (male) obesity is illustrated.

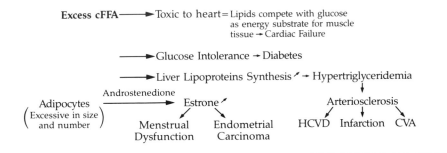

Figure 2–5. Some of the metabolic consequences of obesity.

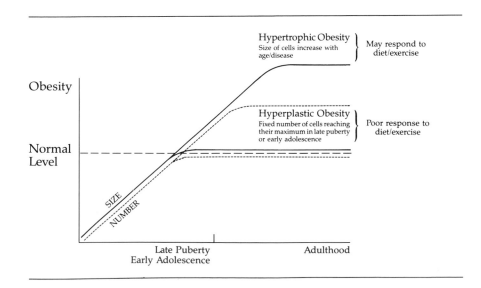

Figure 2–6. The theory of hyperplastic and hypertrophic obesity is diagrammatically represented.

It appears that obesity commencing during childhood is caused by an increase in both the number and size of fat cells and is referred to as *hyperplastic obesity*, although both hypertrophy and hyperplasia occur. Obesity that commences after full development is achieved, that is, during adulthood, is attributable to an increase in the size of the cells and is referred to as *hypertrophic obesity*.[21,30,36,37] These two types of obesity differ in their response to diet and exercise. The hypertrophic type will respond more readily to diet and exercise, whereas the hyperplastic type will not.[31] The presence of two catecholamine-sensitive receptors—beta-1, which promotes lipolysis, and alpha-2, which inhibits lipolysis—will further determine which deposits of body fat will be more active and apt to diminish with diet or exercise and which will be resistant to such activities (Fig. 2–6).[38,39]

Although we are born "programmed" for a predetermined number of adipocytes, and adult fat cells are known to be incapable of multiplication, other factors may activate residual reticuloendothelial cells or primitive fat organs into developing new adipocytes. This seems to be the case particularly in marked obesity, in which the excess number of fat cells may stimulate this reservoir metabolically into forming new fat cells through an adipogenic factor.[40-43] Conversely, the loss of weight experienced by some patients undergoing liposuction surgery may be attributable to a negative feedback system through a neuroendocrine mechanism as if a built-in thermostat in adipose tissue is set at a point at which fat storage (and consequently body weight) is diminished (see Fig. 2–1).[44]

The removal of some of these cells from a fixed number of localized genetically determined fat cells results in a permanent change in the body's shape. By contrast, diet will cause a diminution in the size of the fat cells and not in their number. The difficulty in maintaining a lower weight after a weight-reducing diet may arise because the same number of fat cells is present both before and after the diet, and the activity of the adipogenic factor is determined by the number of adipocytes and not their size.

In addition, certain brain peptides, such as beta-lipoprotein and beta-endorphin, are found in greater quantity in the pituitary gland and plasma of experimentally obese animals and may be instrumental in the development of obesity. It is not clear, however, whether these peptides control the appetite or the metabolism.[45]

METABOLISM

Fifty years after Virchow suggested that fat cells are specialized cells, Von Gierke found evidence that the adipose tissue has its own internal metabolism.[46] The early 1900s marked the beginning of an intense study of adipocytes, including their embryonic development, genetic characteristics, and endocrine regulation. It took another 50 years to develop the technique of calculating the number of fat cells in the body. Pre-adolescent obesity was confirmed as both hypertrophic (caused by increase in size) and hyperplastic (caused by increase in number), whereas adult obesity is mainly hypertrophic. Because the adult adipocytes cannot multiply, their number remains fixed.[47,48]

In addition to providing thermoregulation

through their insulating ability, fat cells influence water balance by their ability to bind it. It is the storage and release of energy, however, that profoundly influences the metabolism of the entire body. Although the internal metabolism of fat cells is free of hormonal control, catecholamines and insulin have a profound effect on the metabolism of the cell.[49,50]

Catecholamines are the most effective lipolytic agents. They are instrumental in providing energy by breaking down the stored triglycerides into free fatty acids (FFA) and glycerol. This occurs mainly by the binding of the catecholamine to the beta-1 receptor present on the surface of the adipocyte wall. The nucleoprotein formed stimulates the enzyme adenylate cyclase to mediate the transformation of adenosine triphosphate (ATP) into cyclic adenosine monophosphate (cAMP), which controls the rate of lipolysis in the human body by its effect on lipoprotein lipase.

This breaks down the triglycerides into FFA and glycerol (G) (Fig. 2–7).

The opposite occurs when the catecholamines are bound to the alpha-2 receptors. This inhibits the adenylate cyclase and consequently the formation of cAMP. The lipoprotein lipase cannot be formed, and the converse action takes place with increased glucose uptake and lipid synthesis.[51–53]

It is possible that differential concentration of the two types of catecholamine-sensitive receptors, alpha-2 and beta-1, accounts for the difficulty or the ease with which fat is lost from different areas of the body. An increase of the antilipolytic alpha-2 receptors and insulin receptors in excessive abdominal and trochanteric fat certainly explains why diet or exercise may not be successful (Fig. 2–8).

Furthermore, the antilipolytic effect is more marked in the femoral area, where even during fast-

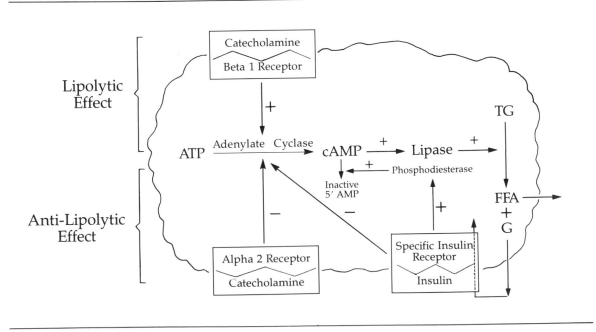

Figure 2–7. The lipolytic and antilipolytic biochemical and hormonal reactions within the adipocyte are outlined.

ing the cells maintain their size despite increased breakdown of triglycerides. A further indication that the femoral cells are richer in antilipolytic receptors is the fact that the catecholamines are more active in the abdominal area (possibly because of more beta-1 receptors), whereas insulin has a strong antilipolytic effect on the femoral area, possibly because of a greater number of specific insulin receptors on those adipocytes.[54]

Adipose Tissue

Abdominal & Trochanteric "Genetic Fat"	Facial, Arms Upper Torso
Rich in Alpha 2 & Insulin Receptors (Antilipolytic Action)	Rich in Beta 1 Receptors (Lipolytic Action)
Poor Response to Diet and/or Exercise	Good Response to Diet and/or Exercise

Figure 2–8. The basic differential localization of alpha-2 and beta-1 receptors in the human body.

Insulin also exerts its antilipolytic effect by binding to a specific adipose cell receptor. From there, through various mechanisms that have not been completely elucidated, it inhibits lipolysis by inhibiting adenylate cyclase and consequently the lipoprotein lipase. This action results in increased uptake of glucose and the formation of intracellular triglycerides. Insulin also stimulates phosphodiesterase, which converts cAMP (which activates the intracellular lipase) into 5'AMP, an inactive metabolite. These activities produce an antilipolytic effect.[55–58]

In obesity and diabetes, the action of insulin is more intricate, and it is possible that only the metabolism of glucose is affected. Furthermore, lipolysis is increased, resulting in elevated circulating FFA, which may cause glucose intolerance which may be harmful to the heart, and may increase triglycerides, which in turn may cause arteriosclerosis.[39,59,60]

It has certainly been proved that obesity is significantly associated with cardiovascular and metabolic disease. Thus, hypertension, cerebrovascular damage, coronary heart disease (angina pectoris, cardiac failure, and infarction), diabetes mellitus, amenorrhea, and a host of other metabolic problems are associated with obesity (see Fig. 2–5).

The previously described independent circulation of fat lobules, when associated with an excessive amount of fat, imposes an additional strain on the cardiovascular system. This may be critical when the system is already impaired.

With obesity affecting approximately one fourth of the population of the United States, control of the condition would considerably reduce morbidity and mortality, decreasing the incidence of coronary heart disease, cardiac failure, or brain infarction by approximately one third. This is an awesome figure, considering the amount of personal devastation rendered by these maladies, not to mention the billions of dollars spent in medical care for a condition that, in principle, could be controlled.[34,35]

Although in women gynecoid-type obesity has considerably less medical implications, general obesity extracts its toll. The main source of circulating estrogen (as estrone) in postmenopausal women is the adipose tissue. Obesity increases its secretion, and there is evidence that endometrial cancer is related to a higher level of estrogen.

Furthermore, obese women tend to have a higher incidence of menstrual dysfunction and hirsutism. It is apparent that the neuroendocrine mechanism with its intricate axis influences not only the formation of adipose tissue but metabolic diseases associated with its excess as well.[61,62]

The difficulty associated with research in this field is evidenced by contradictory findings: although it has been shown that only pituitary extract has an adipogenic effect, sera from hypophysectiomized animals still exhibit this activity.[47] Nevertheless, further research in lipid metabolism has been stimulated by the relatively new field of liposuction surgery, which can remove large quantities of genetically determined adipose tissue and can potentially have a profound effect on metabolic diseases associated with an excessive number of fat cells.

Recent developments in immunology may bring about new ways to combat obesity. It is conceptually possible to block alpha-2 receptors, which have an antilipolytic effect. This action will enable the beta-1 receptors to exert their lipolytic effect in certain areas of the body in which weight loss is particularly difficult to achieve. Carrying this further, obesity itself may be controllable by manipulation of the different receptors on the adipocytes.

Principles of Modern Liposuction Surgery

3

The success of modern liposuction technique is based on several factors, some intrinsic and some extrinsic.

INTRINSIC FACTORS

Primitive Fat Organs

The location as well as the density of the primitive fat organs are genetically determined and differ from person to person. Thus, the location of fat is genetically determined.[22–28]

Hyperplastic and Hypertrophic Obesity

The facts that adipocytes do not duplicate themselves and that by adulthood one will have the number of cells that will be present throughout life are discussed in Chapter 2 (see Fig. 2–6). Thus, cells removed will not return, but those left behind can hypertrophy concomitantly with the rest of the fat cells present elsewhere.[36–42] Liposuction is particularly effective in areas in which fat cells are rich in antilipolytic alpha-2 receptors. These areas, such as thighs and abdomen, contain genetically determined fat, giving an inherited shape to the person; they do not respond well to diet or exercise (see Fig. 2–8).

EXTRINSIC FACTORS

Instrumentation

Cannula. The progression from a sharp curette to a blunt-tip cannula with side openings led to the evolution of liposuction from a possible procedure to a successful procedure. In addition to the blunt tip of the cannula, a more aerodynamic tip (e.g., a bullet shape) causes less trauma to the essential tissue traversing the fat, such as blood vessels, nerves, and fibrous septa (Fig. 3–1).[6,8,9,63]

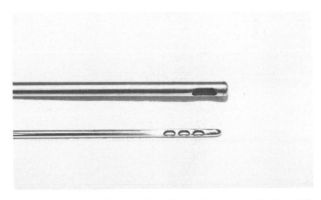

Figure 3–1. An early cannula of large diameter with blunt tip and one aperture is compared with a modern cannula with small-diameter bullet-shaped tip with three apertures.

The use of smaller-diameter cannulae over the past few years also contributed to the improved results of liposuction. These good results are achieved by a network of tunnels at different levels of the adipose layer. The smaller the diameter of the cannula, the more layers of these interconnected tunnels that can be created (Fig. 3–2). This will diminish the possibility of surface irregularities when fibrosis takes place within each channel left by a larger aspirating cannula (Fig. 3–3).

Suction Apparatus. A basic knowledge of the atmospheric pressure as related to the maximum vacuum pressure created within the aspirating unit is necessary to understand its effect on the technique of liposuction surgery and on the fat itself.

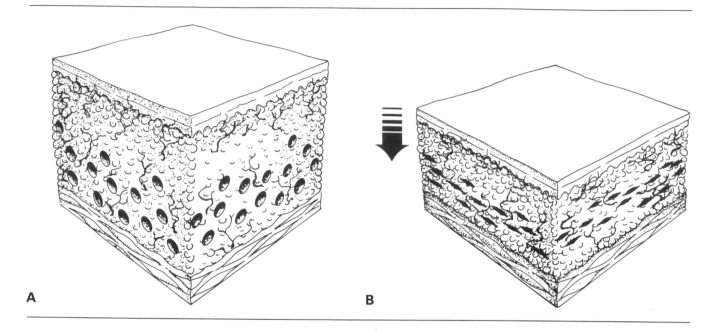

Figure 3–2. A small-diameter cannula will create several layers of tunnels with less likelihood of surface irregularities after the pressure dressing is applied.

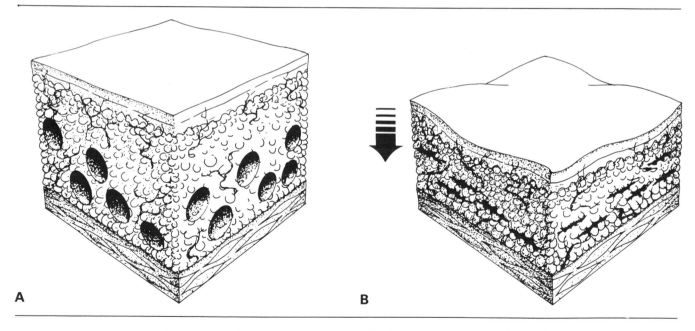

Figure 3–3. A large-diameter cannula is more likely to create postoperative surface irregularities because the tunneling cannot be layered as with a smaller cannula.

The pressure of the atmosphere against the surface of the earth is an effect of the earth's gravitational pull. It has been measured by placing an inverted sealed tube containing mercury in a basin that also contains mercury. Thus, the pressure exerted by the mercury in the tube against the mercury in the basin equals the pressure exerted by the air against the mercury in the basin itself (Fig. 3–4).

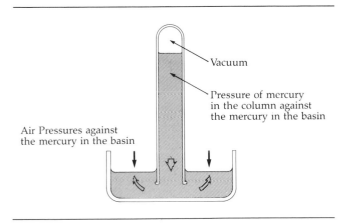

Figure 3–4. Torricelli's barometer diagrammatically illustrates the equilibrium established between the pressure of the mercury in the column and the atmospheric pressure against the mercury in the basin.

The pressure exerted by the weight of the atmosphere varies at different altitudes. At sea level, the height of the mercury in the tube is 760 mm, or 29.9 inches. At higher altitudes, there is less pressure on the mercury in the basin. Consequently, the mercury in the tube will meet less resistance in its

downward force, more of the mercury will pass into the basin, and the height of the column will diminish.

In a suction unit, a vacuum is formed by the removal of all air (rarely attainable) from an enclosed system, whereby the pressure inside the system is below the atmospheric pressure. At higher altitudes, there is less pressure per unit area. Consequently, an aspirating unit gauge may show a lower value at a higher altitude, even though the negative pressure or the vacuum created within the closed system equals that at sea level.

The difference in the reading of the gauge is the difference in atmospheric pressure between sea level and the altitude at which the new reading is taken. For example, a reading of 700 mm Hg at an atmospheric pressure of 760 mm Hg is equivalent to a reading of 640 mg Hg at an atmospheric pressure of 700 mg Hg; that is, the vacuum pressure is the same in both cases because of the air still present within the system, which is the same in both cases.

By understanding the rudiments of physics, one can better understand the effect on suctioned fat of an aspirating unit using different types of cannulae. With a closed system, in which the cannula is within the fat and there are no air leaks, the fat is suctioned through the aperture of the cannula into the cannula itself and from there through the connecting tube to the collecting container, which has a negative pressure (i.e., a vacuum). The resistance of the fat globules still attached to the surrounding fat has to be overcome by the vacuum pressure, necessitating a higher pressure. A lower pressure would accomplish the same results but would be more traumatic to the fat because it would tear it away from the surrounding tissue. This would mean not only more trauma (with more bleeding) but also a longer time to perform the surgery.

By using a curette-cannula—a cannula with sharp-edged apertures—the fat globules are first cut away from the surrounding tissue and then aspirated into the tube. Thus, the fat suctioned into the aperture does not offer the resistance of fat globules that are still attached. A lower pressure can be used for its removal because the aspiration is of free fat (Fig. 3–5).

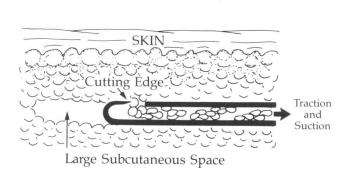

Figure 3–5. According to the principle of Kesselring's original curette, a true suction lipectomy is performed by shaving off layers of fat and suctioning them into the cannula.

Although Kesselring understood the physics of liposuction, his assertion that there is no difference between the sharp and blunt instruments, and that one half atmospheric pressure of vacuum force is sufficient for liposuction is puzzling. Certainly loose fat globules produced by curetting can be suctioned at a lower pressure than can attached globules.[64,65] Furthermore, the negative pressure required to suction the fat is inversely proportional to the diameter of the cannula and the size of its aperture. Thus, the larger the diameter, the less vacuum pressure required.

In addition, the creation of air leaks by withdrawing the cannula close to the surface or using a cannula with a built-in air line will enable the suctioned air to push the aspirated fat forcefully along the tube and into the collecting container; this happens every time the cannula is removed with the vacuum on.

Some South American surgeons have designed their aspirating cannulae, blunt and sharp, with an air line that permits the surrounding atmospheric pressure to enter the tube, thereby taking advantage of lower pressure in the suction unit.[66,67]

The optimum pressure to aspirate fat with a closed system seems to be as close to a full atmospheric pressure as the unit will allow.[8,68,69] When a high pressure is reached within the suction unit, vaporization of the fluids present within the suctioned material occurs because the atmospheric pressure within the tube and collecting jar is significantly lower[67]; that is, there is less air weight to contain the material intact. An analogy would be the action of a child's balloon, which pops as it rises in the atmosphere because the pressure around it is not sufficient to maintain the shape and volume of the balloon.

To elaborate, at sea level (760 mg HG atmospheric pressure) water vaporizes (boils) at 100C. The lower the atmospheric pressure, the lower the temperature at which vaporization takes place. Consequently, by lowering the pressure in the aspirating unit, vaporization will take place at a much lower temperature than 100C. This vaporization of liquid is caused by increased molecular movement of the liquid, which physically aids the movement of the fat globules along the aspirating tube.[70] This effect facilitates the surgery by speeding up the procedure, thereby diminishing the operating time. The less time the patient is on the operating table, the less medication will be needed (e.g., sedation, analgesia).

Nevertheless, because the effect of vaporization can extend to the aspirated fat globules, this is not the preferred technique when the removed fat is intended for reinjection, as in autologous fat transplantation.

The injection of chilled saline solution into the fat prior to liposuction has been found to facilitate the procedure considerably. It is indeed possible that the lower temperature of the aspirated liquid aids in its vaporization.

Technique

New techniques in modern liposuction have made the procedure safer and more successful.

Tunneling. Kesselring described his technique of creating a large space within the panniculus adiposus, which, by compression, becomes a virtual space. His technique consisted of introducing the sharp curette-cannula close to the fascia with the aperture *upward* toward the fat and skin. By withdrawing the cannula and exerting pressure upward, he was literally shaving off layers of fat that were aspirated in the cannula (Fig. 3–5). By repeating this process continuously, he performed a "continuous, or layered, suction lipectomy," in the belief that this method achieved a smoother skin surface. However, he reported fairly frequent seromas.[7,64]

Modern technique uses the concept of tunneling at different levels and in different areas. With the use of smaller cannulae, the tunneling could be performed at different levels and achieve a smoother effect than if performed with large cannulae.[8,9,71,72] Fischer strongly advocates vertical tunneling at different levels with some distal overlapping, to avoid hanging of the skin (due to gravity), which would be increased by horizontal tunneling (Fig. 3–6). In addition, vertical tunneling also decreases the possibility of excessive blood loss.[73–76]

Crisscross Technique. Fournier further modified the procedure by creating a network of tunnels perpendicular to each other (crisscross technique), a technique that was apparently first performed by Fischer[6] (Figs. 3–7, 3–8) and later by R. L. Dolsky (personal communication, 1986). This is achieved by making several incisions and having the tunnels interconnect from different directions creating a honeycomb or swiss cheese effect.[77–81] Comparing the fat removal area with a sponge better illustrates the

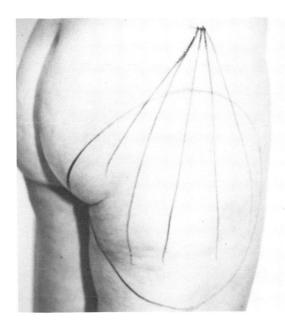

Figure 3–6. Vertical tunneling, as advocated by Fischer, avoids horizontal ridging caused by gravity on the skin.

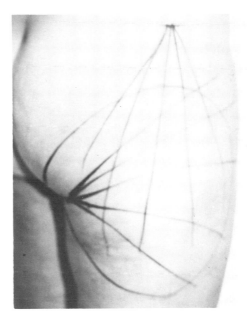

Figure 3–7. The crisscross technique of tunneling is advocated by Fournier.

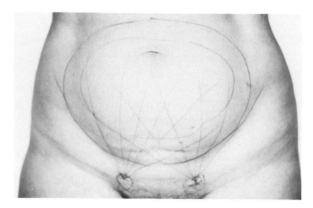

Figure 3–8. Multiple entries can create a finer network of multi-layered interconnecting channels.

Figure 3–9. The concept of the spongelike morphology of the fat after liposuction is demonstrated. Note the numerous interconnecting septae between the deeper and the more superficial layers of fat.

Figure 3–10. The compressive dressing will obliterate the spaces left by liposuction tunneling. The contraction of scar tissue within each tunnel will maintain the desired shape after healing is complete.

spaces left by the removal of fat from different directions as well as the interconnecting tissue (containing intact nerves and blood vessels) between the fascia and the skin (Figs. 3–9, 3–10).[82]

Wet and Dry Techniques. Another principle on which modern liposuction surgery is based, especially when performed under local anesthesia, is the wet technique. Injection of a liquid in the adipose tissue to be removed (an anesthetic, a hypotonic solution, or just a carrier for epinephrine or hyaluronidase) characterizes the wet technique.[83] This technique, which can vary with the amount of the fluid injected, is discussed in greater detail in Chapter 4.

Tape or Garment. Although this approach is disputed by some, taping the abdomen following liposuction redrapes the skin into a position that, when healed, will give the desired cosmetic result. Taping the operated area is particularly important when liposuction is performed in more than one stage to correct a large defect. This subject is also discussed in greater detail in Chapter 15. It should be emphasized, however, that no garment can exert the pressure that well-applied tape can exert on an operated area.

In conclusion, modern liposuction technique takes advantage of several factors:

1. The genetic location of fat attributable to the primitive fat organs.
2. Fat cells do not reproduce themselves, which led to the theory of hyperplastic and hypertrophic obesity.
3. The type of cannula used (i.e., blunt-tip) is more aerodynamic in design, with a smaller diameter and smoothly polished apertures.
4. A suction apparatus can reach nearly one atmospheric negative pressure in its vacuum.
5. The technique of tunneling in different directions and at different levels is made easier by the use of a smaller-diameter cannula.
6. The wet technique is certainly the technique employed when local anesthesia is administered; it facilitates liposuction and causes less trauma and bleeding.
7. Tape is used to remodel the area operated on because it exerts greater pressure than a garment.

The Wet and Dry Techniques of Liposuction Surgery

4

The terms *wet* and *dry* refer to injection of a solution into the adipose tissue to be aspirated. The wet technique can be used with or without general or epidural anesthesia, whereas the dry technique can be used with either general or epidural anesthesia.

The original technique used by Illouz, to which he still adheres, includes general anesthesia as well as injection of a hypotonic solution, with or without epinephrine, to which hyaluronidase is added as a spreading factor.[8] In the belief that the hypotonicity of the solution helps lyse the adipocytes, Illouz employed the term *lypolysis* to describe his technique.[84,85] This term is still used by many general plastic surgeons in the United States.

Hyaluronidase was used to aid in the diffusion of the hypotonic solution which, in principle, helped emulsify fat cells, facilitating their removal. The present use of hyaluronidase is discussed in Chapter 10.

Although Illouz used general anesthesia for analgesia, he employed the wet technique to facilitate the removal of fat and diminish the blood loss. Thus, the initial definition of wet technique included general anesthesia as well as injection of a solution.[86] Fournier modified the procedure by employing only general anesthesia and not using any injectable solution[77,80,87]; he then began using a solution containing epinephrine for vasoconstriction. His more recent use of plain, chilled saline solution injected with fat prior to liposuction (known as neocryoanesthesia)[88,89] seems to add further anesthesia and vasoconstriction without addition of any potentially toxic pharmacologic agents.

Fournier's initial dry technique, however, gained the acceptance of most general plastic surgeons. These surgeons introduced it to their hospitals to be performed under general anesthesia, with all the protocol and under all the aseptic conditions with which major surgery is performed in American hospitals.

One of the reasons the dry technique was accepted over the wet one was because of its convenience. It does not require the time-consuming prior injection of a solution into the areas to be operated on.

The current definition of the wet technique includes the use of a regional or general anesthetic with the injection of a saline solution (hypotonic or normal), with or without hyaluronidase and with epinephrine.[90,91]

Cosmetic surgeons who perform surgery in their offices and who do not want to use general anesthesia in a nonhospital setting have further modified the wet technique by the addition of a local anesthetic to the injection solution. These surgeons used this modified wet technique from the very beginning, which was only later described by Hetter[92] and by Fournier as the new "wet" technique.[71] Thus, liposuction surgery performed under local anesthesia is, by definition, performed with the wet technique. The injection of a local anesthetic containing epinephrine prior to liposuction surgery performed under general anesthesia is now done by many surgeons to diminish not only the blood loss but also the morbidity of the procedure and the postoperative discomfort.

Personal experience, as well as observation of the two techniques as performed by various cosmetic surgeons, led to the conclusion that there is far less blood loss with the wet than with the dry technique despite variations in instrumentation (Fig. 4–1).

Some critics of the wet technique have cited the possible distortion of the area to be treated because of the prior injection of the solution and the subsequent difficulty in the quantification of the fat extracted. In addition, they complain that the process of the injection itself is time consuming. This distor-

Figure 4–1. This 1100 ml of fat practically devoid of blood was aspirated from a patient's abdominal wall using the wet technique.

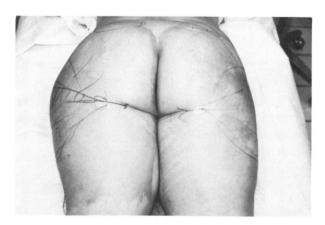

Figure 4–2. This intraoperative photograph demonstrates the use of the wet technique. The right side has been infiltrated, and liposuction was performed on it. The left side has not been infiltrated.

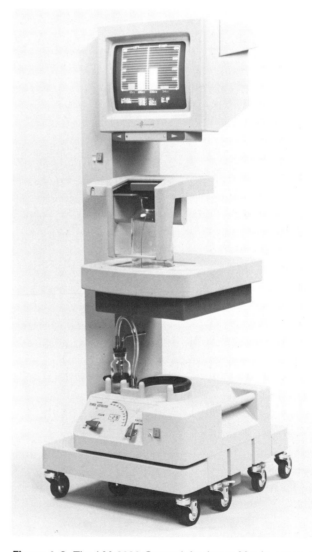

Figure 4–3. The LM 2000 General Aspirator Monitor uses electronic technology to measure separately the tissue removed and the accompanying fluid loss. *(Courtesy of M.D. Engineering, Foster City, California.)*

TABLE 4–1. COMPARISON BETWEEN THE WET AND DRY LIPOSUCTION SURGERY TECHNIQUES

	Dry	Wet
Distortion	No	Yes
Ease of extraction	Less	More
Time consumption	No	Yes
Blood loss	More	Less
Potential		
Seroma	More	Less
Hematoma	More	Less
Ecchymoses	More	Less
Fatocrit*	±20%	±10%

*A relative measurement of blood-content in the extracted fat.

tion has not been found to be of significance, and the advantages of the wet technique far surpass the possible disadvantages (Fig. 4–2). The advantages and disadvantages of the two techniques are summarized in Table 4–1.[91]

Other surgeons have also found the addition of epinephrine to the injected solution to diminish the blood content in the extracted fat from 20 to 25 percent to 5 to 10 percent.[93]

Using the same technique, Dolsky found that the drop in hematocrit correlates with the quantity of fat extracted. There is a 2 percent hematocrit drop when up to 500 ml of fat is extracted; this increases to 10 percent when 1000 ml of fat is removed. Thus, the drop in hematocrit increases as the quantity of extracted fat is increased.[94] It must be remembered, however, that the hematocrit determination may not be an accurate indication of blood loss because it can be changed by the amount of intravenous fluids given pre-, intra-, and postoperatively since that can cause a relative hemodilution. Similarly, the shifting of intravascular fluid into the extravascular space may result in a relative hemoconcentration. More recent work indicates that loss of blood, contrary to its benign appearance in the fat-collecting container, is more than it appears.[95] Although some surgeons suggest that 2000 ml of the aspirate should be the limit in an outpatient facility,[186] none have used the wet technique as outlined in this textbook or in my previously published manual on liposuction under local anesthesia. Through its use, one can safely aspirate considerably more than this arbitrary volume. Calculation of true blood loss is useless if one does not differentiate clearly between the different protocols used. A new aspirator has the ability to electronically monitor and graphically display the quantity of tissue removed and the accompanying fluid loss (Fig. 4–3); it may also be helpful in these calculations.

Indications for Liposuction

5

Liposuction is not a treatment for obesity, but it is eminently effective in removing unwanted localized deposits of fat that are genetically determined and that cannot be lost by dieting. It is well known that men and women tend to accumulate fat in different parts of the body (android and gynecoid types of obesity), sometimes giving rise to a true deformity or lipodystrophy. The configuration of lipodystrophies differs not only in men and women but also in people from different backgrounds. Thus, Caucasians differ from Asians, who have a different morphology from persons of African origin.[48,72]

The success of liposuction is a function of the inability of fat cells to reproduce themselves; there is evidence that the hypothalamus acts as a body thermostat by negative feedback.[44] Once a certain number of fat cells are removed from areas in which fat metabolism is slow (presumably those areas containing alpha-2 receptors that inhibit lipolysis), the cells do not return, and the entire organism may adjust itself to a reduced requirement of lipids. This may be one explanation as to why some patients who undergo liposuction either maintain their weight or lose it easier than they did before surgery.

Although the ideal candidate is usually a person under the age of 40, it is the physiologic rather than the chronologic age that may be important.[71,96,97] This depends on the patient's health, muscle, and skin tone. Localized accumulations of fat on the face, arms, breasts, chest, abdomen, thighs, hips, and legs are all amenable to liposuction surgery.

In women, excessive fat deposits localized in the trochanteric area are the main complaint; consequently, this is the most frequently performed surgery (Fig. 5–1). This genetically determined fat localized on thighs is extremely resistant to diet and exercise, and almost every patient requesting its removal by liposuction surgery has tried numerous diets and exercise regimens without success. Liposuction on the hips is followed in frequency by the abdomen (first lower then upper), the neck and jowls, inner thighs, knees, arms, calves, and lastly, the ankles.

The most common areas men want corrected with liposuction surgery are the abdomen (Fig. 5–2) (first upper then lower) and the flanks (Fig. 5–3), the so-called love handles. Enlarged breasts caused by true or pseudo-gynecomastia are the third area that many men want corrected (Fig. 5–4). It must be real-

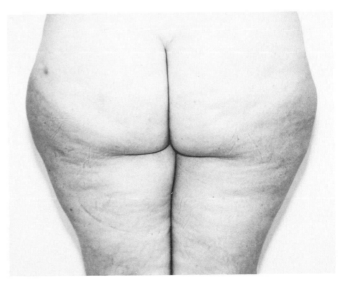

Figure 5–1. Excessive trochanteric, genetically determined fat does not respond to diet or exercise.

ized, however, that much of the fat that gives men the protuberant upper abdomen may be caused by omental and not subcutaneous fat.

Although obesity is not an indication for liposuction, South American surgeons have achieved considerable success in treating it with this method. Liposuction can be safely used in the obese for body contouring and not for weight reduction.[98] There is also evidence that its use as a staged procedure may be useful for induction of weight loss as well as for the patient's appearance (Figs. 5–5, 5–6).[83,99–102]

The use of liposuction in certain metabolic diseases is under investigation.[103] Liposuction has been known to reduce the insulin requirement in diabetics. Reducing the thickness of the adipose layer in a patient whose cardiac function is impaired may be beneficial in reducing the peripheral circulation; the excess fat can place an additional stress on the heart by increasing its work load.

In addition, it has been used successfully in reconstructive surgery. Thus, defatting of flaps,[104,105]

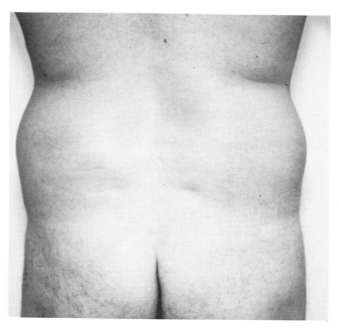

Figure 5–3. Excessive fat is deposited in the flank area ("love handles") in this 38-year-old man.

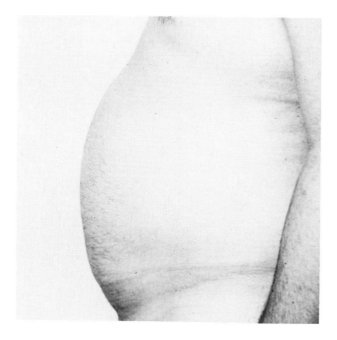

Figure 5–2. Abdominal fat deposit is shown in a 42-year-old man.

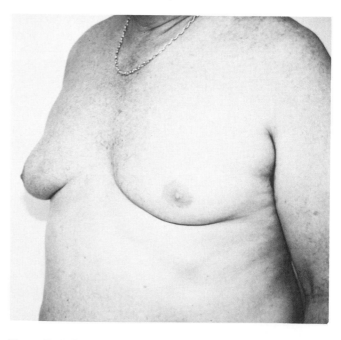

Figure 5–4. Gynecomastia associated with moderate obesity is shown in this 55-year-old man.

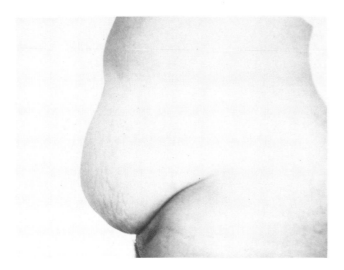

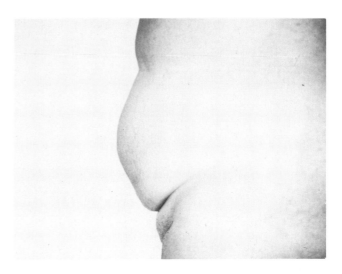

Figure 5–5. This preoperative photograph is of an obese patient with a moderately pendulous abdomen. The patient chose to have liposuction done in stages rather than undergo an abdominoplasty.

Figure 5–6. Compare the postoperative view of the same patient shown in Figure 5–5 after the first stage of liposuction. A minimum of 3 months should be the interval between procedures.

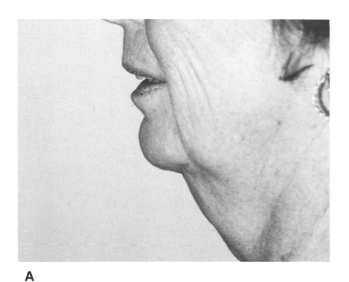

A

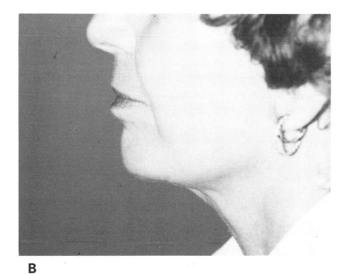

B

Figure 5–7. Compare these pre- and postoperative photographs of a patient who underwent rhytidectomy with liposuction as an adjunct procedure (see Chapter 14).

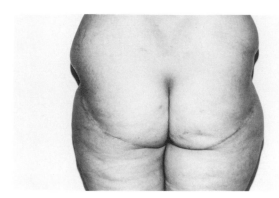

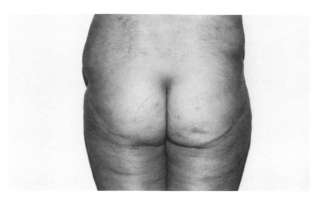

Figure 5–8. This body deformity followed a thigh lift performed by a plastic surgeon.

Figure 5–9. The deformity shown in Figure 5-8 is corrected with liposuction surgery. Should the patient desire it, a touch-up can be performed within a few months.

lymphangioma circumscriptum,[106] lipomatosis of the neck (Madelung's or Launois-Bensaude disease),[72,107,108] and morbid obesity involving the neck[109] have all been treated succesfully with liposuction. Similarly, the technique of mobilizing a large flap by performing soft tissue dissection with a blunt-tip liposuction cannula has been used successfully as an adjunctive procedure in both cosmetic and reconstructive surgery.[110,111] Congenital lymphedema has also been treated with liposuction surgery.[48,112]

Liposuction has been used successfully as an adjunct procedure to certain cosmetic procedures, such as rhytidectomy[83,113–117] (Fig. 5–7) abdominoplasty,[118–121] medial thigh "lift,"[122,123] and breast reduction.[124] Liposuction can also be helpful in correcting body deformities caused by plastic surgery (Figs. 5–8, 5–9).

Extraction of buccal fat pad with liposuction has proved a safe and effective procedure, when properly performed, for subtle reduction of the fullness of the cheek.[125,126] In addition to aesthetic facial and body contouring, liposuction can be used for the removal of lipomas, which may be either moderate or huge in size (Fig. 5–10).[97,127,128]

The clinical safety of liposuction, if properly performed, has been established for more than a decade[129]; recent microscopic studies of removed adipose tissue confirmed that liposuction removes only fat cells: "not once have parts of the supporting connec-

tive tissue been found."[130] It is this supportive tissue that carries vessels and nerves.

In conclusion, liposuction is indicated for a variety of conditions (medical, cosmetic, and reconstructive), assuming that both physical and mental health of the patient are good. One of the most difficult tasks for a cosmetic surgeon is learning to say "no." Good clinical judgment as well as experience and maturity will bring about better results if strict criteria are used in choosing candidates for surgery. The retractability of the skin appears to be the most important factor in accepting or rejecting a patient.

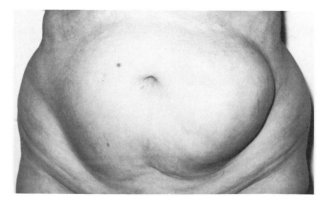

Figure 5–10. A huge abdominal lipoma in a 68-year-old woman.

Equipment

6

The equipment necessary for liposuction surgery consists of a cannula, used to create tunnels inside the fat layer, and an aspirating unit strong enough to suction the fat out.

THE CANNULA

The operator is clearly more important than the operating instrument. It is also true that better instruments facilitate the surgery and increase its safety. The first instrument used for liposuction was the abortion curette. This was found unsuitable for routine use particularly because of its sharp edges and flexibility. Fischer's cannula operated by first suctioning the fat into the aperture and then cutting it by means of a sharp moving blade. The outside design of Giorgio Fischer's cannula is closest to the one used today with best results, that is, an aerodynamic bullet-shaped tip with an aperture approximately $\frac{1}{2}$ inch from the tip. Without the sharp blade, it represents an ideal instrument except for its small opening.[6] Schrudde[2] and later Kesselring[7,131] used sharp aspirating curettes that basically severed the fat and then suctioned it. Still later, Teimourian used a modified fascia lata stripper based on the same design principle as Schrudde's and Kesselring's.[68] However, his initial technique had an approximately 30 percent incidence of seromas.[69]

There are numerous modifications to the cannula design, all based on Illouz's initial instrument. Some have one, two, or three apertures on one side only, whereas others have five: one ventrally (the side of the cannula that almost always should be kept away from the surface), and two on each side laterally. The larger the aperture and the more apertures present (to a degree), the more efficient the instrument becomes (Figs. 6–1, 6–2).

Elam's development of the cobra-tip cannula represented a true innovation (Fig. 6–3). The design of the instrument made possible liposuction of areas in which dense fibrotic tissue was present, such as in true or pseudo-gynecomastia in men. It is no longer necessary to perform disfiguring incisions around the areola to correct this embarrassing condition.[132]

It is true that the "power stroke of the cobra cannula occurs when it is pushed"; that is, the fat and fibrous tissue are broken down and aspirated as the instrument is introduced into the fatty tissue.[132] The power stroke of the traditional cannula, however, is present both when the instrument is introduced and when it is withdrawn, because it has more aperture surface exposed to the surrounding fat.

The cobra cannula is an efficient instrument that can be dangerous in the hands of an inexperienced operator. The safety of the procedure is limited when the tip of this cannula is close to the skin. It can also break more blood vessels (with more subsequent bleeding) because of its head-on encounter with the vasculature present in the adipose layer. It is this head-on encounter that makes the instrument so effective in correcting gynecomastia through a very small incision.

Several variations of Elam's cannula manufactured by different companies have recently appeared on the market. It must be emphasized that the safety and efficiency of the instrument depend more on the hand of the operator than on its design. While the cobra cannula as well as Illouz's blunt-tip instrument can be dangerous in the hands of an inexperienced operator, both can be safe when used by an experienced physician. The margin of safety of the cobra-design cannula, however, is much smaller than that of the traditional blunt-tip cannula.

In addition to the different cannula tips, the shape of the cannula tube may vary, some having angulations and some having curvatures in different directions as related to the site of the apertures. The

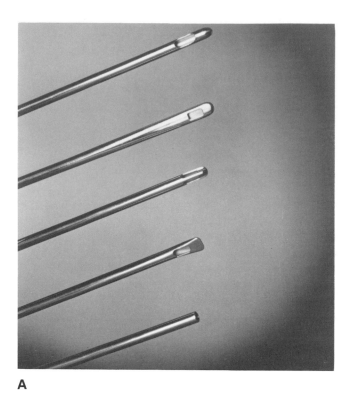

A

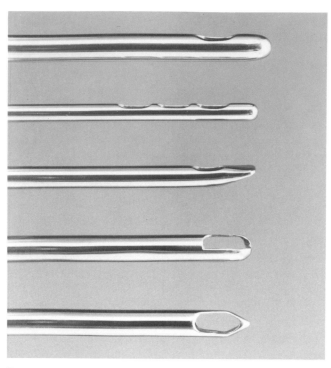

B

Figure 6–1. Various cannulae demonstrate the different types of tips. *(Courtesy of Reliance Medical Corp., Grand Junction, Colo.)*

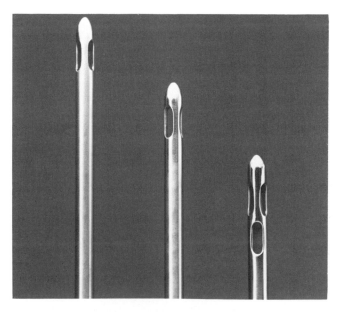

Figure 6–2. Cannulae with different number of apertures. *(Courtesy of Byron Medical, Tuscon, Ariz.)*

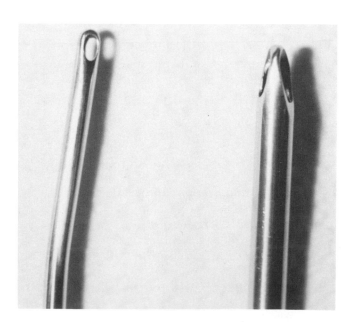

Figure 6–3. Elam's cobra cannula *(manufactured by Wells-Johnson Co.)* is particularly useful for correction of gynecomastia.

diameter may also vary, from 2 to 3 mm (used in facial lipo-sculpturing) to 8 to 10 mm (used primarily for body liposuction). The most common cannula used has a 6 mm diameter (Figs. 6–4 through 6–6).

The handle of the cannula also varies from manufacturer to manufacturer. A good grip on the handle is essential, and the round smooth handles are not as popular as those that have a rough-textured surface, preventing the hand from slipping (Figs. 6–7, 6–8). In addition to the texture of the handle, its design also varies with some being round, some hexagonal, and some molded, to offer a better grip. Furthermore, the handle can be in line with the cannula, at an angle (pistol-grip)[133] or perpendicular to the cannula. The pistol-grip cannula works well in areas in which the hand and handle do not impinge on the adjacent anatomic structures. If the handle is turned 180 degrees so that it is opposite the aperture, the angle of the grip becomes uncomfortable, against the natural position of the grasping hand. The cannula with a perpendicular handle is manufactured so that the aperture is opposite the handle, which is held similar to a saw. (Fig. 6–9).

The suggested cannulae are 3 mm and 4 mm short (15 to 20 cm long) and 4 mm, 5 mm, and 6 mm long (25 to 28 cm long). There may be a personal preference for one design over another, depending on training, feel of the instrument in the hand, and cost. However, the safest cannula to start with is one with a blunt-tip bullet-shape end with a distal aperture 8 to 10 mm from its tip. Having two or three apertures, the efficiency of the instrument is increased. The edges of these openings should be smoothly polished so that they will not act as a curette while the instrument is moved back and forth under the skin. Instruments with different length, diameter, and curvature may be acquired, depending on the operator's personal preference and experience (Fig. 6–10).

THE ASPIRATING UNIT

During the past few years, a plethora of suction machines have appeared on the market. Each manufacturer proclaims the advantage of his unit over the others (Figs. 6–11 through 6–13).

Certain facts should be kept in mind when purchasing such a unit:

1. The suction apparatus should reach the desired vacuum pressure (approximately 760 mm Hg) in a relatively short time (i.e., 5 to 15 seconds).
2. There should be enough residual vacuum left in the tube when the cannula is removed and reinserted immediately so that the operator is not slowed down by waiting for the pressure to rise again.
3. The unit should be easily serviceable either by the manufacturer or any other knowledgeable person.

When the purchase of a suction device is considered, one may want to compare all these factors as well as price, particularly because a backup unit should be available whenever liposuction is performed. The malfunction of a suction unit in the middle of a procedure can be extremely disturbing to the patient and the physician alike.

There are certain advantages to having a unit with two visible collecting containers: the operator can extract more fat than with a single collecting jar without having to change it, and the quantity and quality of fat extracted from one side into one jar, with the fat extracted from the other side into the other jar, can be visually compared (Fig. 6–14).

Unless one intends to perform liposuction only on the face and neck, a high-power suction unit is necessary. It increases both the efficiency of the surgery and the safety of the patient by shortening the operating time.

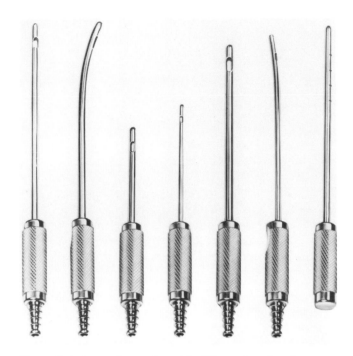

Figure 6–4. A set of cannulae with different lengths, diameters, and curvatures. *(Courtesy of Padgett Instruments, Kansas City, Mo.)*

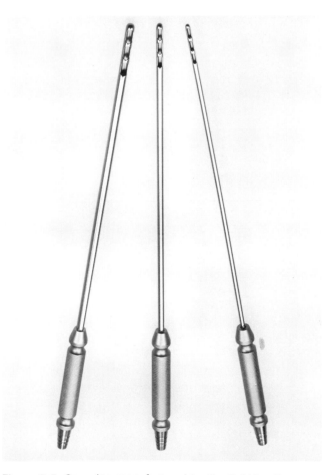

Figure 6–5. Cannulae manufactured by the Robbins Company are shown. The efficiency of the instrument is increased by having more than one aperture.

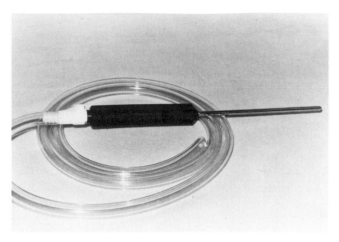

Figure 6–6. A cannula manufactured by Dunhill Medical.

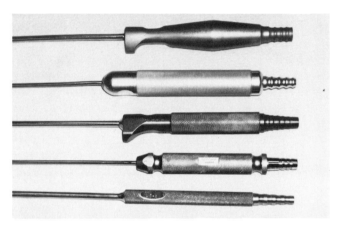

Figure 6–7. Different types of handles of liposuction cannulae.

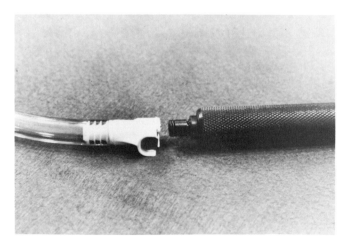

Figure 6–8. This closeup of the Dunhill cannula handle shows the easy-on, easy-off adapter.

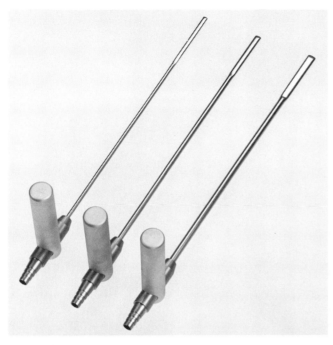

Figure 6-9. The Robbins' Paragon cannulae with the handle perpendicular to the suction tube is flattened near its tip. The aperture is opposite the handle.

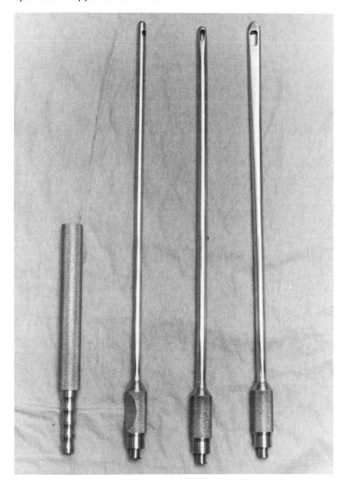

Figure 6-10. The Newman Quick Connect Cannula System with the universal handle is shown with the bivalve, cobra, and spatula cannulae. *(Courtesy of Grad Tech, Philadelphia.)*

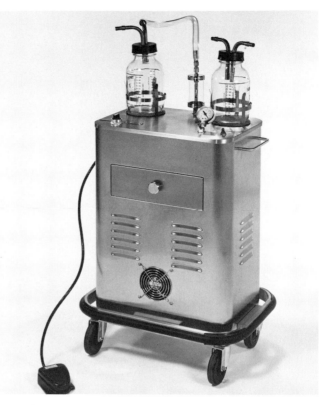

Figure 6-11. The Robbins' aspirating unit was one of the earliest in the United States.

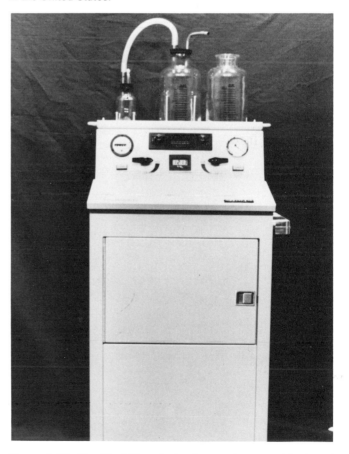

Figure 6-12. The Dunhill aspirator has a storage cabinet and even a radio incorporated within the unit.

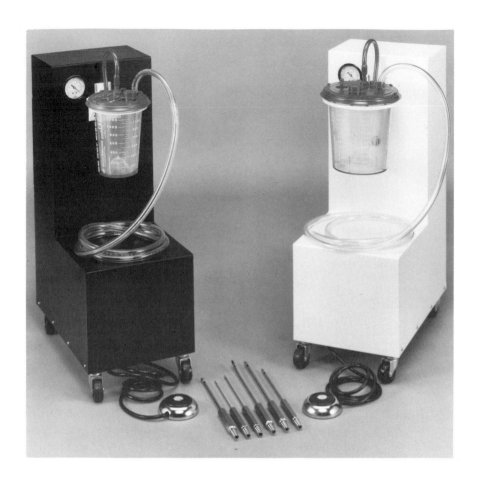

Figure 6–13. The Dean Medical aspirator has disposable collecting containers.

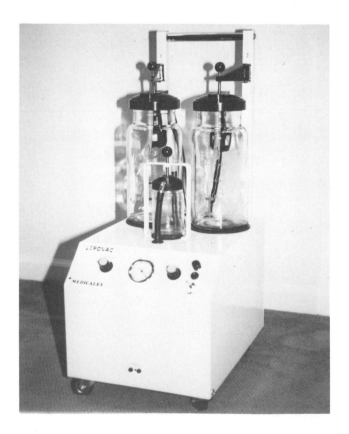

Figure 6–14. The original French aspirator, which is now available with two collecting jars, rather than only one.

Initial Consultation

7

From the surgeon's point of view, a procedure may be a technical and aesthetic success, but if the patient is not happy with the results, the surgery is a failure. That patient will let his or her dissatisfaction be known to every friend and acquaintance, and the surgeon will lose many potential patients.

During the initial consultation, not only is the physician evaluating the patient, but the patient is evaluating the physician. The physician's confidence and technical knowledge of the procedure can be easily detected by a perceptive potential patient, particularly by a well-informed one. Public lectures on cosmetic surgery, particularly on liposuction, which is a relatively new procedure, will acquaint prospective patients with this method of removing fat and will enable them to hear questions asked by other persons (Fig. 7–1).

After this initial contact is made—when quite often the first and lasting impressions are formed by both the patient and the doctor—the physician's assistant should take the patient to the examining area. There, the patient changes into a garment appropriate for the examination.

The patient's initial consultation for liposuction, as for any other surgical procedure, elective or medically required, represents a two-way street. The physician must establish rapport with the patient, which is essential for the patient's satisfaction and consequently for the success of the procedure.

Evaluation should be performed in a systematic manner, with the patient standing on a stool, preferably before a full-length mirror.

PATIENT'S COMPLAINT

Physical Complaint

The patient should point out the defects to be corrected. The first question, often: "What can you do for me, doctor?," should be answered with another question: "What bothers you the most, what are your priorities?" Only after the essential complaint is voiced by the patient should the surgeon make other suggestions complementary to it. Other areas to be corrected, as suggested by the physician, will usually give a more total improvement in the patient's appearance. For instance, if the patient wants her thighs reduced and she has very prominent hips, this fact should be pointed out to her. Reducing the thighs will correct that defect, but the total shape will not be as pleasing without correcting the hips as well.

Psychological Complaint

The patient's complaint should be evaluated in light of the importance the patient attaches to the relative "deformity." The psychological evaluation of the patient should be continued throughout the examination. Often the patient's complaint about an

Figure 7–1. Participants are shown at a public lecture on liposuction, held at the author's surgical center.

33

undesirable feature of the body may reflect a more serious complaint or a dissatisfaction with life. Dissatisfaction with one part of the body is common. When many areas are involved, however, and the patient wants the entire body "done over," the patient may have serious problems that may be psychological or related to domestic or work difficulties. Similarly, if too much importance (a relative term in view of the different attitudes with which the patient and the physician view the problem) is given to a relatively small defect, the surgeon should be wary of making promises to the patient. Correcting a small defect will, by necessity, bring a modest improvement, and the patient should fully understand that the change will be a subtle one.

Liposuction is not a treatment for cellulite—dimples and irregularities in the skin. If present before the surgery, they will certainly be present afterward. Realistic expectations by the patient are most important for the success of all cosmetic procedures (Fig. 7–2).

If a physical complaint can be related to a possible psychological one, the cosmetic surgeon may be able to detect a potentially difficult patient. Some patients do not have realistic expectations for what the surgery will do for them. These patients hope that correcting the physical defects will change their lives.

PHYSICAL EVALUATION

The patient's suitability for the procedure should be evaluated during the initial consultation, and not during subsequent visits. A second visit to the surgeon implies acceptance of the case.

Good results depend on good choices. The physiologic age of the patient may be more important than the chronologic age. Skin laxity and muscle tonus (particularly in the abdominal area) should be

tested and pointed out to the patient. For the different areas of the body that are amenable to liposuction surgery, these factors should be considered separately.

Face and Neck

One of the most frequent complaints by women is the accumulation of fat under the chin. In evaluating this area, several factors should be taken into consideration.

Turgidity of the Skin. When the skin is lax, even if not excessive, liposuction results may not be satisfactory. Laxity of skin without sufficient contractability will result in excessive skin folds. A cervicofacial rhytidectomy may be indicated (Fig. 7–3).

Excessive Skin. The typical "turkey gobbler" configuration of the neck, a complaint voiced especially by

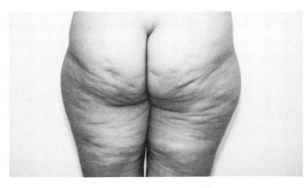

Figure 7–2. Irregularities in the skin's surface (cellulite) should be pointed out to the patient prior to surgery. Liposuction will not remove them and may even make them more pronounced if the tunneling is too close to the skin.

men, is mainly attributable to excessive skin. Liposuction by itself will not correct the problem and may even aggravate it. Here, too, a cervicofacial rhytidectomy may be required.

Muscular Laxity. Often fullness under the chin (which may be mild or enough to cause a midline "wattle") may be caused by muscular laxity. This fullness may be present with or without a collection of fat. Redundant platysma in the midline, as well as fat accumulation between it and the muscles that form the lower aspect of the floor of the mouth, are factors to be considered in examining the undersurface of the chin; removal of fat below the platysma muscle must be performed under direct vision through a submental incision. In younger persons, fullness under the chin may be due to a congenital ptosis of its "strap" muscles. This is accentuated when swallowing and the patient should feel this movement to understand the limitation of liposuction under the chin.

Gynecomastia

Examination of the enlarged breast will usually demonstrate whether the condition is the result of true gynecomastia. If this is the case, the fibrous quality of the glandular tissue will be palpable, particularly under the areola, but also extending into the rest of the breast. In pseudo-gynecomastia, the entire breast tissue is soft (due to the adipose tissue) with only a small nubin of glandular tissue palpable as a disk under the areola. It is much easier to correct pseudo-gynecomastia than true gynecomastia, which has both hypertrophy and hyperplasia of the mammary glands. Correcting true gynecomastia may also require a subareolar incision for excision of these glands, and the patient must be told of this possibility. Excessive gynecomastia may require excisional surgery (Fig. 7–4).

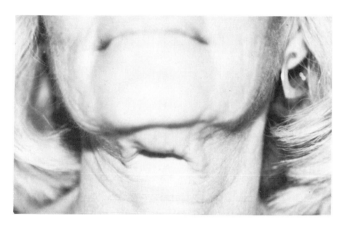

Figure 7–3. Excessive skin and lack of turgidity require a cervicofacial rhytidectomy.

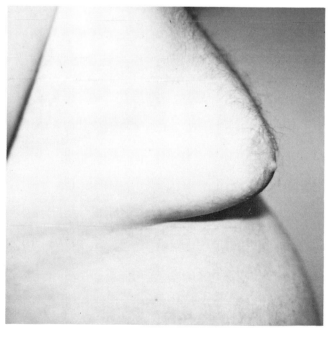

Figure 7–4. Excessive gynecomastia may require either liposuction in stages or excisional surgery.

In addition, hypertrophy of the pectoral muscle often gives the impression of fullness of the breast. The patient should be made aware of this condition by being asked to tighten the chest muscles and feel the overlying fat while the muscle is contracted (Figs. 7–5, 7–6).

Upper Arms

Evaluation of the upper arms for liposuction surgery requires the same circumspection as for calves and ankles. Patients desiring liposuction of the arms (usually the posterior aspect) generally fall into two categories. The first includes young women whose arms are not too heavy, yet they wish them to be thinner. The consistency of fat and laxity of skin is excellent, but the improvement will be subtle and the patient has to understand this thoroughly. The second category includes women who have very heavy upper arms but whose skin is too lax to have a good result from the usual procedure. In this case, the arms form folds that hang when the arms are extended. To visualize the hanging folds better, the patient should raise the arms horizontally and bend the elbows with the forearms upward (Fig. 7–7).

Aside from these two main groups, some women have heavy arms that would respond well to liposuction surgery performed in stages. Careful evaluation of these patients is imperative, and the limitations of this procedure must be particularly well explained.

Upper Abdomen

The adipose tissue in the epigastric area is more fibrous and vascular than in the lower abdomen and quite often feels nodular, particularly in the older patient. This will result in a "cellulite" configuration in the skin with multiple dimples. This also should be pointed out to the patient because quite often the nodularity in this area will be increased by liposuction (see Fig. 7–11).

In men, the supraumbilical area is often very large and may be caused not only by increased subcutaneous fat but also by a thick layer of omental fat and distention of the stomach. Extreme caution should be exercised in promising the male patient a significant reduction of the supraumbilical fat layer; an abdominoplasty may be more indicated (Fig. 7–8).

Lower Abdomen

The subcutaneous fat in the lower abdomen may vary considerably in thickness, and a weak protruding abdominal wall will give the impression of a heavier layer of fat than is actually present. To test the strength of the abdominal musculature, the patient should be placed in a supine position with the legs straight and the feet together. The skin and its subcutaneous fat should be grasped and the patient asked to raise the legs together slowly, keeping them straight. The grasped fat that initially seemed so thick will escape from the operator's fingers and will feel much thinner against the contracted and protruding abdominal wall. The patient should try to

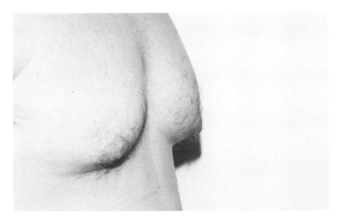

Figure 7–5. A preoperative view of moderate gynecomastia is shown.

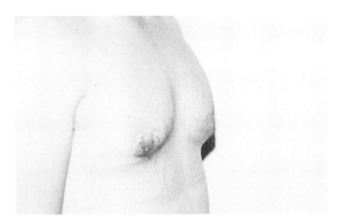

Figure 7–6. This postoperative view shows the pectoral muscle still giving the impression of fullness of the breast.

feel this change in the apparent thickness of the fat to understand that the protruding abdomen is caused not only by fat but by a weakness in the muscle as well.

A patient is more likely to accept a less-than-perfect result of liposuction of the abdomen if the cause of its protrusion is understood before the surgery. The protuberant abdomen may not be completely flattened by liposuction. A diathesis of the rectus abdominalis can also be detected by this manuever; this should also be pointed out to the patient (Fig. 7–9).

An incipient pendulous abdomen can often be corrected with liposuction. Occasionally, it may require more than one procedure (staged liposuction). This must be thoroughly discussed with the patient, who may elect a one-stage abdominoplasty (dermolipectomy) over two or more liposuction procedures (Figs. 7–10, 7–11).

By having the patient pull in the abdomen, stretch marks will become more obvious, and the patient will have a better concept of how he or she will look after surgery. Even if the patient does not ask about the stretch marks, this maneuver should be performed during the initial consultation. The physician should inform the patient that stretch marks will be more obvious after the surgery. An explanation before surgery creates better understanding, whereas an explanation after the surgery often appears as an excuse.

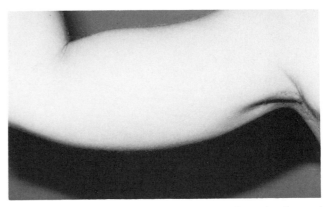

Figure 7–7. The position of the arm visualizes the fat to be removed.

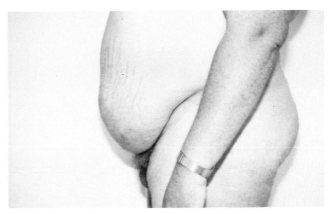

Figure 7–8. Excessive omental fat will not respond to liposuction surgery.

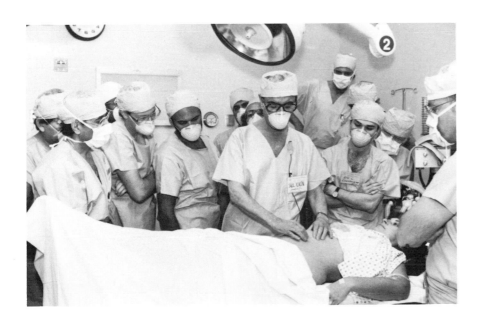

Figure 7–9. Grasping the abdominal wall while the patient raises the legs will accentuate the abdominal protuberance as a result of muscle weakness.

When a patient has a relatively thin subcutaneous adipose layer and wants it thinner, liposuction becomes a procedure that can give equivocal results at best. The thinner the fat layer, the closer the cannula will be to the skin. This is very likely to give an uneven and nodular surface, and the patient should be told explicitly of the possibly questionable results.

Retracted abdominal scars can respond well to liposuction. By removing the excess fat on each side of the scar and freeing the retracting fibrous tissue from under it, the abdominal wall can become smoother and the aesthetic result can be quite pleasing.

An abdomen containing rather soft fat, which has a jellylike movement when pushed in one direction or another (ballottement), will respond less well to liposuction than one with firm fat. Pinching the skin and its subcutaneous fat will give a good indication of the turgidity of the skin and its retractability after the surgery.

Flanks

Excessive fat in the flanks, the so-called love handles, is one of the most common complaints by men. The surgeon should make the patient aware that the anterior aspect of this deposit is over the iliac crest and is the result of increased thickness of skin and fibrous tissue rather than subcutaneous fat (Fig. 7–

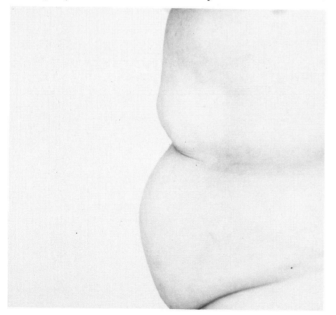

Figure 7–10. This preoperative photograph shows a patient with incipient pendulous abdomen. The patient was told she might need more than one procedure.

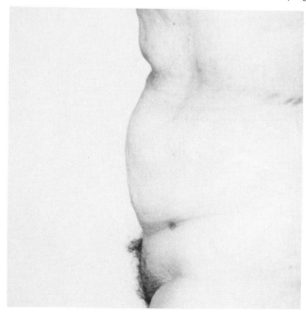

Figure 7–11. This postoperative photograph was taken at 3 weeks. A second procedure may not be necessary for this patient.

12). The posterolateral deposits respond well to liposuction (Fig. 7–13).

Hips

Liposuction of hips usually gives excellent results; the hip is a forgiving area and a good place for the novice to start. This correction is usually performed the same time as the thighs. Some women have an exaggerated flare of the posterolateral aspects of the iliac crest. In this case, liposuction may not achieve the desired result because the large hips are caused by bony formation and not fat accumulation (Fig. 7–14).

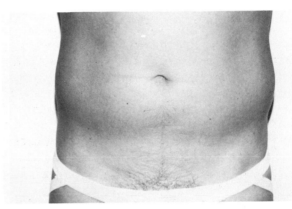

Figure 7–12. The bony prominence of the anterior iliac crest can simulate fat deposits.

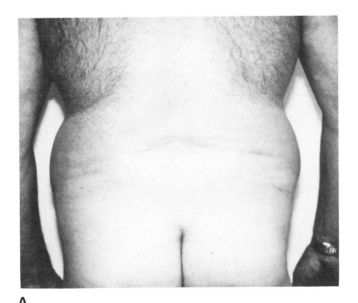

A

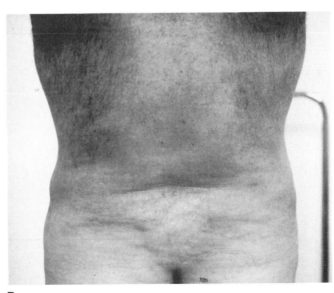

B

Figure 7–13. Pre- and postoperative views of liposuction of flanks (love-handles) are shown in a 32-year-old man.

Thighs

Excessive trochanteric fat, so-called saddle bags or riding breeches, is the most common complaint by women. Often, simple removal of the excess fat from the thighs will not improve the contour to its best, and liposuction of the lateral lower thighs should also be performed. In examining the patient for this common problem, particular attention should be given to the consistency of the fat (ballottement), and the skin should be checked for horizontal drape-like folds. This will denote poor retractability of the skin and this "cascade of waves" may become more apparent after liposuction. If the patient is more concerned with appearance when dressed (rather than in a bathing suit), liposuction will correct the bulge and the patient will be satisfied. Such patients consider these possible defects of the skin's surface rather minor and consider liposuction a small price to pay to have a better shape.

The thighs and buttocks are also areas in which cellulite is commonly present (Fig. 7–15). If present before the surgery, these dimples and skin irregularities will be present after and in some cases may even appear worse. The patient should understand that cellulite is not helped by liposuction. An attempt to eliminate it would mean operating the cannula superficially and not leaving a sufficient layer of fat attached to the skin. This not only interferes with circulation but creates a rather thin skin covering over a layer of fat, which would create a cottage cheese-like surface. The results would obviously be worse than before surgery.

Medial Thighs and Knees

The medial thighs and knees are discussed together because both have an anatomic and aesthetic relationship. Medial thighs are particularly apt to form drapelike folds, particularly in the older patient. The result of liposuction in this area is usually rather subtle, and the patient should understand the minimal changes that this procedure will bring. Often the bulging medial thighs will appear more

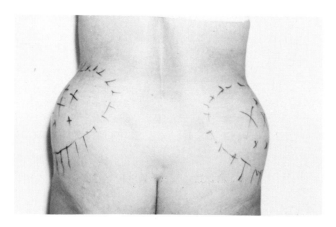

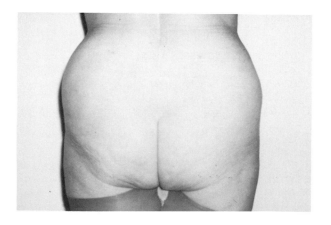

A

B

Figure 7–14. Prominent hips are occasionally the result of pelvic formation, and the response to liposuction will be limited.

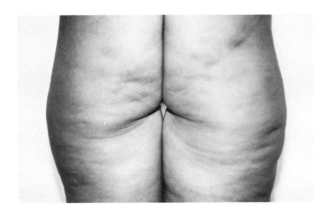

Figure 7–15. Cellulite of the buttocks and thighs does not respond to liposuction.

prominent because of the valley formed between them and the fat deposits on the inner aspects of the knees.

Removing the excess fat from the medial aspect of the knees will often lessen the prominent appearance of the upper medial thighs and will give a pleasing appearance to the lower limbs. Furthermore, removal of the medial protuberance of the upper thighs will create a rectangular appearance, with the perineum forming the upper aspect and the straight vertical medial thighs the two vertical lines of this geometric figure. This can be demonstrated by having the patient stand in front of a full-length mirror. By pulling back the skin of the inner thigh, the surgeon can reproduce this space between the thighs. Diminishing the width of the thighs will improve the general appearance of this area (Figs. 7–16, 7–17).

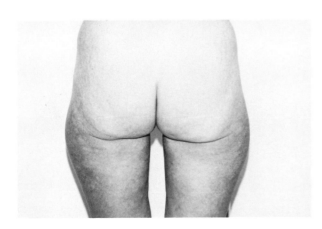

Figure 7–16. This preoperative photograph shows "saddle-bags," accentuating the space between the upper medial thighs.

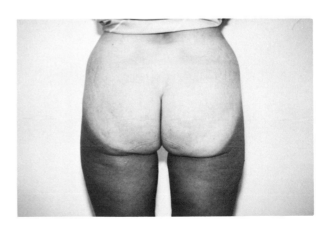

Figure 7–17. This postoperative view shows the improved spatial relationship by removal of the trochanteric fat.

Calves and Ankles

Evaluation for improving the shape of calves and ankles requires a great deal of experience on the part of the surgeon. These areas should be approached with utmost care and circumspection. The technique is very exacting and has no margin for error. Injury to the underlying structures is easy to inflict, and postoperative edema can be prolonged.

If the edema does not subside within a reasonably short time, it can become fibrous. This will require a secondary procedure, which can be more difficult than the first one. Because of the prolonged dependent edema, the patient may not be able to return to work for at least 2 weeks. This differs from liposuction of other areas of the body in which the patient may resume normal activities within a few days of surgery (Fig. 7–18).

CONCLUSION

The initial examination of the patient is most important. It should be performed in an unhurried manner without causing embarrassment to the patient (Fig. 7–19). This problem will be obviated if the back is examined before the patient is asked to turn around to face the physician. The surgeon should have a good grasp of the procedure to be able to make an intelligent decision about what can and what should and should not be done. Asymmetries

should be pointed out to the patient; two Polaroid photographs should be taken, one to be given to the patient and one to be kept with the chart. After the patient is dressed, a frank discussion of the contemplated procedure should follow, aided by Polaroid photographs.

This discussion with the patient should include the following points:

1. The sedation and anesthesia to be used as well as the procedure itself, and the location and size of the incision should be described.
2. Immediate and delayed postoperative care (e.g., bed rest, tape, garment, ultrasound should be discussed.
3. Immediate and delayed results, such as edema following surgery and how long it will take to subside, are to be mentioned.
4. Possible adverse effects should be described. The positive as well as the negative aspects of the procedure are important—the patient must understand its limitations. The surgeon should inform the patient of possible complications while stressing the fact that properly performed liposuction surgery is one of the safest cosmetic procedures today.
5. If the patient is not a good candidate, he or she should be told. One of the most difficult tasks a cosmetic surgeon has is refusing a patient, for whatever reason. It is better to reply with a categorical *no* than with an equivocal *yes*. Whenever

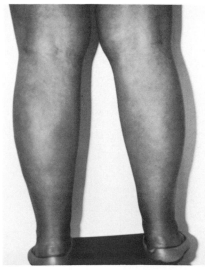

A

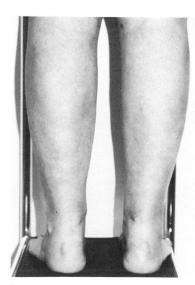

B

Figure 7–18. Compare these pre- and postoperative views of calves and ankles on which liposuction surgery was performed. Post-operative edema can be particularly prolonged in this area.

possible, rejection of the patient as a candidate for liposuction should be accompanied by the offering of an alternative. This can be either another procedure for correction of the defect or a referral to a colleague for consultation. Surgeons must remember that they are above all physicians dealing with patients who have probably thought about the procedure for months, have built it up in their imagination, and have come to the office expecting only positive and encouraging remarks. Most patients desire cosmetic surgery because they want to feel better about themselves; one can usually bear a physical affliction much better than a psychological one.

6. Financial considerations should be discussed freely during the initial consultation. Even if the patient does not bring up the fee, the surgeon should volunteer it as part of the information necessary to the patient.

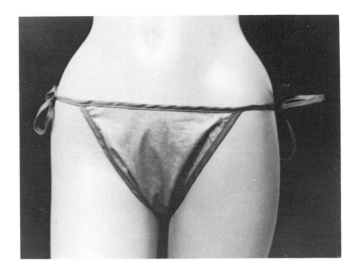

Figure 7–19. A bikini-type garment will ease patient's embarrassment during the initial examination. *(Courtesy of Dunhill Medical, Inc., Jackson, N.J.)*

Presurgical Visit and Postsurgical Instructions

8

It is preferable to have the patient's presurgical visit following the initial consultation at a subsequent visit. The interval between these visits gives the patient time to "digest" the information received during the initial consultation. The better informed the patient is, the better the surgery and the postoperative period will be tolerated. During the second visit, the surgeon will have a chance to answer further questions that clarify certain points and to emphasize once again that the patient should have realistic expectations for the procedure.

PHYSICAL EXAMINATION

A physical examination must be performed to determine the patient's health and suitability for the procedure. Physiologic age is more important than chronologic age. The skin and muscle tonus should be tested and its receptivity to the procedure pointed out to the patient. Good surgical results depend on good judgment by the physician. A history of possible liver disease should be well documented and pertinent blood tests ordered. Impaired liver function can lead to a toxic reaction to the medications used during the operation.[134]

MEDICAL HISTORY

The patient's medical history, medications taken, and untoward results to previous surgery should be noted. Should the patient have a questionable medical history or condition, a letter asserting his or her suitability for the contemplated surgery should be requested from the patient's physician. A verbal approval is not sufficient, and the surgeon should insist that a letter follow the initial consultation.

The position of the patient during surgery should also be determined at this time. Measurement and weight determinations are part of the presurgical visit, as well as fitting the patient for the proper size garment.

CONSENT

Consent for the procedure should be gone over in detail with the patient and signed during the presurgical consultation—not the day of surgery. If there are any doubts concerning the patient's understanding of the consent, the patient should be asked to initial each paragraph as it is read; the patient's companion (and witness) should do the same.

PHOTOGRAPHS

Photographs should be taken from different and reproducible angles. Photographs taken the day of surgery may not be adequate for any number of reasons. Consequently, they should be developed and attached to the patient's chart before the surgery is performed, as they constitute a vital part of the medical records. Photographs should be taken from different directions, with particular attention given to skin defects. These photographs will be the only proof the surgeon has after surgery of any particular defect the patient may not have been aware of previously.

PRESCRIPTIONS

Prescriptions for medication such as vitamin K, vitamin C, antibiotics, and analgesics should be

45

given during the presurgical visit and filled by the patient before surgery. The patient is asked to take vitamin K, 5 mg bid, and vitamin C, 500 mg qid, for 10 days before the surgery. Erythromycin, 250 mg qid, is prescribed for 5 days, starting with the day of surgery.

BLOOD TESTS

A complete battery of blood tests, such as SMAC-24 with a complete blood count (CBC) as well as a full complement of coagulation tests, are indicated.

Other tests may have to be performed as indicated by the patient's medical history.

MEDICAL RECORDS

Medical records made during the initial consultation should be completed during the presurgical visit. This is particularly important when discussing possible complications. The patient's chart should contain details concerning these possibilities in relationship to the specific condition.

Examples of the consent form for liposuction and the instructions given to the patient during the presurgical visit are presented.

Consent for Liposuction

Patient _____ Date _____ Time _____

I hereby authorize _____ and/or his associates to perform liposuction on my _____. I fully understand that this procedure has limited application. No guarantee or assurance has been given to me by anyone as to the results that may be obtained. I am aware that the practice of medicine and surgery is not an exact science. No guarantees or promises have been made to me as to the results of the operation or procedure.

Dr. _____ has discussed in detail with me the information that is briefly summarized below:

I. *Nature and Purpose of Liposuction*
Liposuction is a body-contouring technique. This procedure is a means of reducing localized fat deposits that are difficult or impossible to remove with diet or exercise and is not a treatment for obesity. In liposuction a solution is injected under the skin into the fatty tissue before it is removed. Afterward, the skin is taped, or a girdle is worn for support. Patients usually return to work after 4 to 7 days. Ultrasound and massage may be necessary for several months.

II. *Risks*
I understand that among the known risks are bruising, lumpiness, dimpling, sagging of the skin, scarring, numbness, minor depressions, and periodic swelling of the lower legs, particularly when the procedure is performed on the knees and below them. If skin sagging occurs, a second operation for its excision with additional scarring may be needed. I am aware that in addition to the risks specifically described above, there are other risks, such as loss of blood or infection that may accompany any surgical procedure. Injury to facial nerves, which may lead to temporary numbness or speech difficulty, may occur if surgery is performed on the face. I recognize that during the course of the operation, unforeseen conditions may necessitate additional or different procedures than those described above. I therefore authorize and request that the above-named surgeon, assistants, or designees perform such procedures as are, in the surgeon's professional judgment, necessary and desirable.

III. *Anesthesia*
I understand that local or general anesthesia is normally required when liposuction is performed and that general anesthesia is riskier than local anesthesia. I consent to the administration of local/general anesthesia by or under the direction of _____ . I am aware that risks are involved with the administration of anesthesia, whether local or general, such as allergic or toxic reactions to the anesthetic and cardiac arrest.

IV. *Alternatives to Liposuction*
Alternative methods of body contouring do exist. Although some of them have a longer history, they can leave long scars.

V. *Informed Consent*
I have had sufficient opportunity to discuss my condition and proposed surgery with _____ and all of my questions have been answered to my satisfaction. I believe that I have adequate knowledge on which to base an informed consent to the proposed treatment. No explicit or implicit promises were made to me regarding the final result of the operation.

VI. *Photographs*
I consent to be photographed before, during, and after the treatment and I understand that these photographs will be the property of the above doctor and may be published in scientific journals and/or shown for scientific reasons.

VII. *Cooperation*
I agree to keep Dr. _____ and staff informed of any changes in my permanent address, and I agree to cooperate with them in my aftercare.

(Please initial each paragraph and sign below.)

Patient or Legal Guardian _____
Witness _____
Witness _____

Liposuction Instructions

Please remember that certain surgical procedures performed at our Center are not of any less importance than if performed in the hospital. Your rest and care of the site of surgery are just as important; you should avoid any strenuous activities upon your return home.

I. Before surgery
   A. Have required blood tests at least 2 weeks before surgery.
   B. All prescriptions given by the Physician should be filled 10 days before surgery and taken as instructed.
   C. Arrange in advance all transportation to and from the Center on the day of surgery.
   D. Shower with Hibiclens cleanser the day before and the day of surgery.
   E. Take no aspirin, aspirin-containing medication, or alcoholic beverages for at least 1 week before or after surgery.

II. Day of surgery
   A. Eat lightly and wear comfortable clothes that are easy to take off and put on (a button-down shirt or sweater and no pantyhose).
   B. Please arrive on time; make allowances for traffic delays.
   C. Please shower the morning of surgery; remove contact lenses and *all makeup while at home.*

III. After the surgery
   A. Complete bed rest is helpful for at least 1 day—after that you may be up and around.
   B. Upon your return home, drink lots of fluids, particularly hot broth and mineral water. Take the prescribed medication as indicated.
   C. Bandages and sutures will be removed in 4 to 7 days; elastic bandges or support garments will be applied, after which you should be able to resume limited activities. These will be increased gradually, depending on the location and the extent of the area treated.
   D. Showering and bathing are permitted after the dressings are removed, approximately 1 week after surgery.
   E. Start gentle finger-tip massage (light kneading motion) to operated areas on 10th postoperative day 10 to 20 minutes each area, twice daily.
   F. Exercise may be resumed 2 weeks after surgery. Bicycle riding 10 to 20 minutes per day is recommended (if stationary bicycle is used, do not use any resistance).
   G. Postoperative visits: You will be seen at the Center at frequent intervals for the first few weeks. The exact timing of these visits will vary, depending on the healing process. Please make every attempt to keep these appointments, as it is vitally important that we closely monitor your healing. If you live in a distant city, we prefer that you stay in town for the first few days after surgery.

IV. If your liposuction was under your chin, jowls, or arm, start gentle fingertip massage on the 10th day (twice daily for 10 minutes). The lumpy feeling that can be felt but not seen usually disappears in a few weeks.

V. If you had liposuction of the abdomen
   A. Start sit-ups (with knees bent) 2 weeks after surgery. Do as many as you can, and increase the number daily. Continue for at least 2 months.
   B. Massage the abdomen 15 minutes daily for at least 3 months. It is expected that the abdomen will feel lumpy for several months. (This is normal.) Your support garment should be worn for 4 to 6 weeks.

VI. If you had liposuction of your thighs, start the following exercise 2 weeks after the surgery:
   A. To tighten your thighs and buttocks, kneel with the back straight and palms on your thighs. Lean backward slowly, using thigh muscles. Do not bend the spine, or the tension in your legs will be transferred to the back muscles instead. Hold for a count of 10; then slowly return to starting position. Repeat.

B. For back of thighs; lie down on back, weight on elbows. Bring knee to chest and straighten leg (point toe), keeping it perpendicular to the floor. Repeat five times with each leg.

C. For inner thighs; lie propped on side, raising your top leg as high as you can. Flex ankle (hard), point toe; then lower leg.

VII. Physical therapy following liposuction

The technique of facial and body contouring with liposuction involves the removal of fat from different levels under the skin, thus sculpturing the shape of the body. Following liposuction, compressive dressings are applied, and healing takes place with the body assuming the new shape.

Ultrasound therapy given for several weeks afterward brings down the swelling, helps resolve the bruising, softens the scar tissue, and speeds up the recovery and healing time.

To ensure the best results, it is strongly recommended that ultrasound therapy be followed for 6 weeks, with two to three treatments per week.

Marking the Patient

9

Marking the areas to be operated on is an essential part of liposuction surgery and should be done with the patient standing. This is facilitated by placing the patient on a two-step stool, which is also manufactured with a safety holding bar (Figs. 9–1, 9–2).

Preoperative marking is important because the deposits of fat shift when the patient is lying down. The surgeon will no longer be able to judge the areas of greatest protuberance when the patient moves from a standing to a prone or supine position. To avoid effacing the markings when the patient is prepared with Betadine and Cetylcide, a gentian violet pen resistant to these solutions is employed.*

The marking should indicate not only the areas of greatest protuberance from which most of the fat should be removed, but also the areas in which liposuction should not be performed because of natural indentations or valleys in the skin and subcutaneous fat.

Rotating the patient slightly during this process will demonstrate the edge of the protuberance as well as other irregularities in the skin that may have to be avoided during the surgery. If the photographs taken during the presurgical visit are not completely adequate and do not demonstrate certain skin defects, more photographs should be taken at this time.

Marking is done topographically, starting at the periphery of the fatty deposit with concentric lines toward the crest of the greatest protuberance, which is also marked with an X. A 1- to 3-inch border beyond the deformity is marked with small lines perpendicular to the most peripheral concentric line. This will denote the area in which peripheral mesh undermining or feathering will take place. In order to achieve a smooth transition between the excessive deposit of fat and the surrounding area, the removal of fat should not stop abruptly at its periphery. Instead it should be graded and tapered so that less fat is removed at the periphery than at the center. In addition, these markings will indicate a gradual diminution in the level of liposuction to be performed outside the main lipodystrophic area, sometimes caused simply by creating tunnels beyond it.

When treating the trochanteric area, Fischer marks the patient three different times each with a different color marking pen. His first marking indicates the protruding fatty deposits with the patient standing. His second marking indicates the false protuberances with the patient lying down. These false protuberances are again marked with the patient standing.

The location of the planned incision should also be marked at this time. Figures 9–3 through 9–14 demonstrate markings in different areas of the body.

---

*Richard-Allan Industries, Cat. No. 25010505B.

Figure 9–1. The two-step stool is helpful for examining, photographing, and marking the patient prior to surgery. *(Courtesy of E.F. Brewer Co., Menemonee Falls, Wis.)*

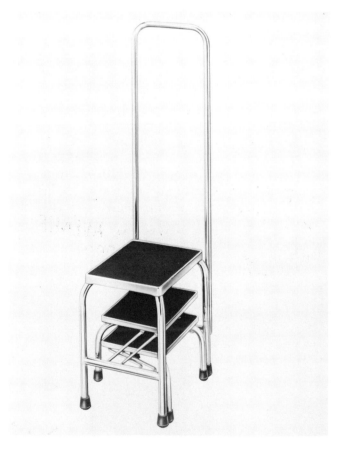

Figure 9–2. The same stool shown in Figure 9–1 is shown folded and with the safety holding rail. *(Courtesy of E.F. Brewer Co., Menemonee Falls, Wis.)*

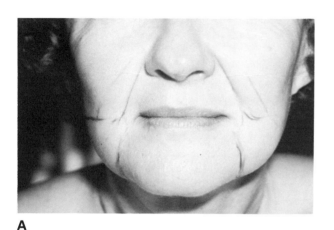
A

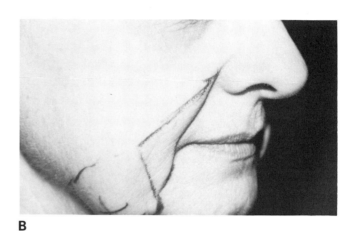
B

Figure 9–3. Markings for liposuction of melolabial folds and jowls.

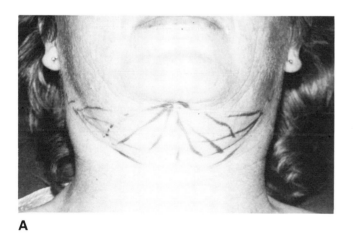
A

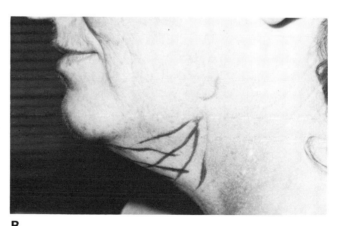
B

Figure 9–4. Liposuction of undersurface of the chin and neck is marked in a crisscross fashion. The three planned incisions in the submentum and lateral neck are visible.

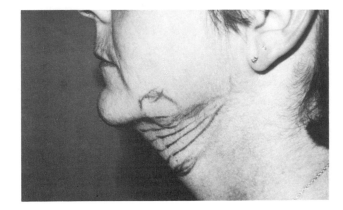

Figure 9–5. The neck and jowls are marked for liposuction using preauricular incisions.

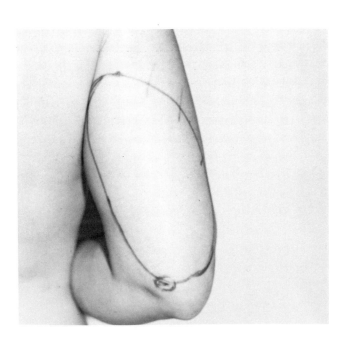

Figure 9–6. The posterior arm is marked for liposuction. Note the additional incision mark on the posteromedial aspect of the elbow.

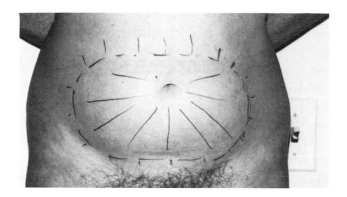

Figure 9–7. Old markings for liposuction of the abdomen are shown. The umbilical incision alone will not remove the peri-umbilical fat properly.

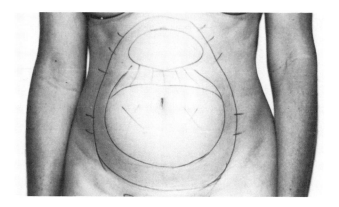

Figure 9–8. The present markings for abdominal liposuction are exhibited. Note the suprapubic incision sites.

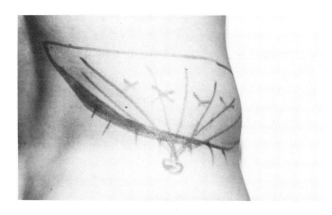

Figure 9–9. Markings for flanks should include the posterior aspect toward the midline. The incision should be made in an area covered by the bathing trunks.

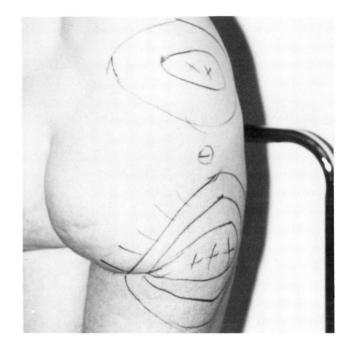

Figure 9–10. Topographic markings for hips and thighs are exhibited. Note the lateral markings on the lower buttocks for feathering. The X's mark the area with the highest protuberance.

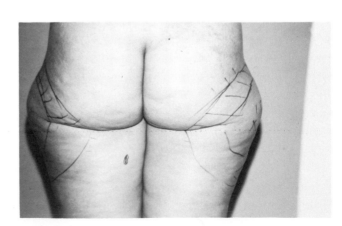

Figure 9–11. Lipodystrophy of trochanteric area shows involvement of the lateral lower buttock.

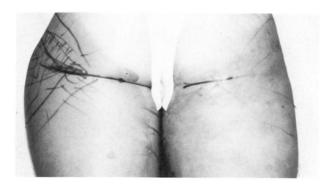

Figure 9–12. This photograph demonstrates the postoperative view of the right side. The left thigh has not yet been infiltrated.

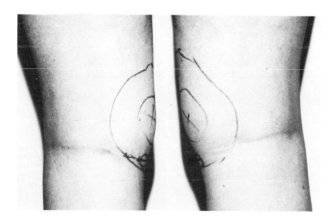

Figure 9–13. The incisions for liposuction of knees should be made posteromedially and within a crease.

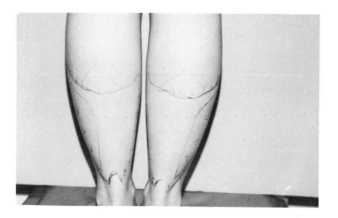

Figure 9–14. Markings for liposuction of calves are displayed. Note the areas of greatest fat deposition (posterolaterally), as well as the incision sites.

Sedation, Analgesia, and Local Anesthesia

10

Over the past 5 years, I have performed thousands of liposuction surgery procedures on the face, trunk, and limbs at a fully equipped outpatient surgical facility. All were performed with the patient receiving intravenous sedation and local anesthesia, with no untoward results and no need for blood transfusion or artificial colloid administration.

The procedure performed under local anesthesia at an outpatient facility has obvious advantages, as well as some disadvantages. General anesthesia has significant risks, not to mention the discomfort and nausea many patients feel afterward. The problem is eliminated when local anesthesia is used, but the use of a fairly large quantity of local anesthetic can have its own risks; these are discussed later in this chapter.

An office equipped as an outpatient surgical facility can be more conducive to the patient's peace of mind (Fig. 10–1). The surroundings are familiar, and by the time the surgery is performed the patient has probably developed a rapport with the office nurses and assistants. This advantage can be countered by the lack of auxiliary personnel and facilities present in a hospital setting. Nevertheless, it is apparent that the recovery time and the rate of infection are the same whether surgery takes place in an outpatient facility or in a hospital.[135]

PHARMACOLOGIC ACTION OF MEDICATIONS USED

Although the surgeon is expected to be thoroughly familiar with the medications used, a brief description of the pharmacologic action of drugs used in liposuction surgery is worthwhile. These drugs include diazepam, naloxone, midazolam, fentanyl, lidocaine, and hyaluronidase. Specific dosages have been specifically omitted, so that the surgeon will do the necessary research to become acquainted with these medications.

Sedation

Preoperative sedation is usually indicated, particularly when the patient exhibits anxiety concerning the surgery. In this case, the preoperative period extends to the time the local anesthesia is administered. This can be either uncomfortable or painful, depending on both the operator's skill and the patient's threshold for pain.

Diazepam (Valium). In liposuction, as with any surgery performed under local anesthesia, amnesia regarding the administration of anesthesia and subsequent procedure is highly desirable. Diazepam seems to fulfill the goal of providing both sedation and a degree of amnesia.

A benzodiazepine with an imidazole ring structure, diazepam is probably one of the most popular nonnarcotic sedatives that can be given either orally or parenterally.

Oral administration results in peak plasma level in less than 1 hr, and in considerably less time when it is administered sublingually because of its partial absorption through the sublingual venous plexus. Intravenous injection results in a much more rapid effect although, because of intimal irritation caused by the propylene glycol present in the vehicle, it can be rather painful. This discomfort can be prevented by intravenous injection of 1 cc lidocaine (Xylocaine) 1 percent plain just before the administration of diazepam.

There is a marked difference in clinical response to the same dose.[136] Consequently, diazepam has to be titrated according to the patient's response; it must not be forgotten that the half-life of the drug is one of the longest of all drugs with similar effects.[137] This results in a longer recovery period, particularly regarding coordination and reactive ability,[138] which may be a detriment when the drug is used on an outpatient basis.[139] Its inhibitory effect on the central nervous system (CNS) seems to be caused by the hyperpolarization of the neuronal synapses.[140]

Although by itself diazepam has few effects on the cardiovascular system, when used with fentanyl it can cause cardiovascular depression, resulting in lower cardiac output, decreased vascular resistance, and hypotension. This has not been found to occur with the technique described in this textbook. A possible reason may be the epinephrine present in the solution containing the local anesthetic. The injected epinephrine may compensate for the decreased plasma epinephrine because of the sympatholytic effect of diazepam–fentanyl combination.[141,142]

Midazolam (Versed). Midazolam is also a benzodiazepine that can be administered either orally, intravenously, or intramuscularly. Because it is more water soluble than diazepam and the commercial preparation does not contain propylene glycol, it is less painful when administered either intravenously or intramuscularly. Its rapid metabolism in the liver gives it a high clearance rate, 10 times faster than that of diazepam.[143,144]

Amnesia following administration of midazolam (anterograde amnesia) seems to be greater than with diazepam.[145,146] This may be beneficial in view of the extensive administration of local anesthesia before surgery.[147] In addition, midazolam causes less postoperative sedation than does diazepam.[148] Thus, midazolam appears to be faster acting, of shorter

duration, and less painful in administration than diazepam.[149]

Analgesia

Although sedation and local anesthesia may be sufficient for performing liposuction, the addition of an analgesic such as fentanyl is helpful, particularly before the administration of local anesthesia. It also complements the anesthetic effect of Xylocaine and has its own sedative effect.

Fentanyl (Sublimase). Fentanyl is a synthetic opioid compound (narcotic analgesic) resembling morphine. This phenylpiperidine derivative resembles meperidine (Demerol). Its analgesic effect is closely related to its stereochemical structure,[150] and it acts through its affinity to specific opiate receptors present throughout the entire brain,[151] and in particular the gray matter and areas involving pain.[152]

Patient response to intravenous fentanyl can vary considerably and may be affected by smoking and alcohol and caffeine consumption.[153] In addition, the response may vary during surgery, as an acute tolerance, even though its plasma level at that time may be higher than when surgery was started.[154] Although adverse cardiovascular effects of fentanyl are rare,[155] they can occur; careful monitoring of the patient is mandatory.

Analgesia can occur with 2 to 10 μg/kg or approximately 0.5 mg or 1 ml administered for a person weighing 140 pounds. Anesthesia, however, may require 10 times that much, which is certainly not advisable in an outpatient surgical facility. The concomitant use of diazepam may increase the possibility of hypotensive episodes.

Bradycardia may occur, although further administration of fentanyl may cause less bradycardia than the initial doses.[156,157] Should it occur, it may respond to atropine administration, which induces a pharmacologic vagal block.

Fentanyl may cause respiratory depression; this effect is dose related, although it is also related to the rapidity of its administration. This is particularly likely to occur when it is given during induction of general anesthesia.[158,159]

Muscle rigidity has been reported with as little as 0.5 ml IV, particularly of the thoracic and abdominal musculature. This rigidity may be related to the rate of administration and is more common in the older patient and when a higher dose is used.[160–162] Diazepam does not by itself counteract the fentanyl-induced chest rigidity.[142]

Whereas seizures caused by fentanyl are rare, nausea can occur frequently even in analgesic doses and can be counteracted by naloxone.

Naloxone (Narcan). Naloxone is an opioid antagonist that acts by competing for the same receptor site as fantanyl.[163] It can be used to reverse the adverse effects of fentanyl, such as excessive sedation, hypotension, chest rigidity, respiratory depression, and any other untoward reactions induced by fentanyl. Its usual dose is 0.1 to 0.2 mg given slowly intravenously. Administration can be repeated every few minutes until reversal of the opioid effects.[164]

Administration of naloxone is not without possible serious consequences, such as tachycardia and hypotension, pulmonary edema, cardiac arrhythmia, and cardiac arrest.[165–167] The sudden reversal of narcotic depression may induce nausea and vomiting.

The rapid reversal of untoward opioid effects induced by naloxone may give the physician a false sense of security. Its duration of action is much shorter than that of the narcotics it acts on.

Consequently, after the naloxone effect wears off, respiratory depression or chest rigidity may reoccur due to renarcotization or to the prolonged effect of fentanyl.[168] Patients requiring naloxone should be observed for several hours after its administration.

Local Anesthesia

Assuming it is practical, anesthesia of the area to be operated upon by infiltration with a local anesthetic has more advantages than disadvantages. The procedure is simple, has fewer postoperative complications, and is certainly less costly than using general anesthesia. The discomfort caused by its administration can be diminished considerably by using oral or intravenous sedation (diazepam or midazolam) associated with intravenous analgesia (fentanyl).

It must be emphasized that local anesthesia as referred to in this textbook implies the use of epinephrine as a vasoconstrictor. It is the addition of epinephrine to lidocaine that makes liposuction under local anesthesia preferable to liposuction under general anesthesia.

Lidocaine (Xylocaine). Local anesthetics induce a lack of sensation in a given area of the body by inhibiting the initiation or excitation process and conduction process in peripheral nerves. The inhibition of certain ionic interactions (particularly of sodium and potassium) through the neuronal membranes does not permit the membrane the necessary changes for the nerve impulse to take place.[169] Thus, the initiation, excitation, and conduction processes are affected by the lack of depolarization of the membrane. New evidence suggests the existence of specific receptors responding to local anesthetics.[170]

Lidocaine is classified as an amide compound because an intermediate chain in its molecule contains an amide linkage, rather than an ester linkage (as is present in another group of local anesthetics).

This linkage, whether amide of ester, contributes to the anesthetic quality of the compound. The duration of the anesthesia can be prolonged by 50 percent by the addition of a vasoconstrictor, preventing the rapid absorption of the anesthetic by the local vasculature.

Lidocaine can have adverse effects on the CNS, such as difficulty in speech, shivering, and tinnitus. Severe toxicity may lead to seizures and CNS depression. Its toxic threshold is increased by 50 percent if diazepam is administered prior to local anesthesia.[171] This is not the case, however, with the cardiovascular toxicity of lidocaine. The potential toxicity of lidocaine is also reduced by the addition of epinephrine, which can lower the venous absorption of the anesthetic by one third.[172] Convulsions can lead to hypoxia of the CNS and should be treated vigorously with 100 percent oxygen.

Since lidocaine is metabolized by the liver, abnormal function of that organ may delay breakdown of the drug and may lead to toxic effects.[173]

True allergy to lidocaine is rare and should be treated accordingly with epinephrine, steroids, antihistamines, and so forth.

Injected lidocaine is rapidly distributed, depending on the vascularity of the area in which it is stored. Thus, it is particularly bound to the fatty storage area.[174]

Hyaluronidase (Wydase). Hyaluronic acid is an essential component of connective tissue ground substance. Hyaluronidase is a polysaccharide that acts as a mucolytic enzyme on hyaluronic acid by causing its hydrolysis. This action decreases the viscosity of the cellular cement and promotes the diffusion, in this case, of Xylocaine and epinephrine.

The increased permeability of the connective tissue caused by hyaluronidase permits 50 percent greater spread of the local anesthetic than without it. This increased diffusion speeds the onset of anesthesia because of the increased distribution of Xylocaine to the nerve fibers. The vasoconstriction induced by the added epinephrine delays the absorption of hyaluronidase into the vascular system, increasing its spreading effect. This vasoconstriction also diminishes blood loss during this procedure (see Chapter 4, under Wet and Dry Techniques of Liposuction Surgery).

Because hyaluronidase is an animal product (extracted from purified bovine testicular hyaluronidase) it has antigenic properties; the patient should be tested for possible allergy to it. A history of allergic reaction to bee stings is an apparent contraindication to its use. Its spreading action can exacerbate a local infection and facilitate rapid absorption of the added epinephrine, with possible untoward systemic effects. Nevertheless, the added epinephrine will induce vasoconstriction, which may slow this effect.

Although the usual dose of hyaluronidase is 1 U/ml of injected solution, as little as 150 units will increase the spread and consequent absorption of 1000 ml of solution.

Allergic reaction to hyaluronidase is very rare and may manifest as urticaria. Symptoms of overdose include nausea, dizziness, tachycardia, and hypotension, which should be treated accordingly.

Hyaluronidase has also been found to help hemorrhagic shock by improving the defective interstitial tissue transport that occurs simultaneously.[175,176]

ADMINISTRATION OF SEDATION AND LOCAL ANESTHESIA

Sedation

To relieve the patient's anxiety, a 5-mg tablet of diazepam is administered approximately 1 hr before surgery. The tablet can either be swallowed or, if a more rapid effect is desired, placed sublingually.

After the patient is taken to the operating room (Fig. 10–2), the area or areas to be operated on are marked. The patient is then placed on the operating table and the cardiac monitor and any noninvasive automatic blood pressure monitor are attached. An intravenous infusion is started using an intravenous catheter, and 2.5 to 5.0 mg diazepam is slowly administered through the IV tubing using a 1-ml tuberculin syringe with a 25-gauge needle. The drip of the IV infusion should be rather slow until the actual surgery commences. This will avoid a quick filling of the patient's bladder; it is rather disconcerting to have the patient want to void in the middle of the procedure.

After preparation of the desired areas, the patient is draped in the usual manner. At this time, 0.5 to 1.0 ml fentanyl is slowly administered IV, also using a 1-ml tuberculin syringe with a 25-gauge needle. A too-rapid injection of fentanyl can cause rigidity of the chest musculature with consequent impairment of gas exchanges. A vial of naloxone hydrochloride (Narcan) should always be on hand in the event that rapid reversal of the fentanyl effect is necessary (Fig. 10–3). The injection of fentanyl is followed by administration of the local anesthetic.

Local Anesthesia

Local anesthesia consists of 50 ml lidocaine 2 percent with epinephrine 1:100,000 added to a 500 ml bottle of sterile normal saline, generally used in intravenous infusion. The use of hyaluronidase, as initially proposed by Illouz, has been fraught with controversy ever since the procedure was introduced in the United States.[92] After repeated and careful comparisons of its use (on one side of a patient as compared with the contralateral side), I have found it useful in reducing blood loss. This is particularly evident if one waits 15 minutes before starting the surgical procedure. Hyaluronidase seems to aid the spread not only of Xylocaine but of epinephrine as well, thus achieving a more even vasoconstriction with a consequent diminution of blood loss. Waiting 15 minutes after the injection of the local anesthetic solution without hyaluronidase did not give the same vasoconstriction and anesthetic effect. With hyaluronidase, the quantity and concentration of Xylocaine and epinephrine in the final injected solution can be cut by one half as compared with the concentration used initially.[83]

If the use of more fluid is anticipated, two bottles are prepared. One may have to remove 50 to 75 ml of saline from the bottle in order for it to accommodate the lidocaine solution. The addition of the local anesthetic solution to the saline bottle is facilitated by inserting an 18-gauge needle (which serves as a vent) obliquely through the rubber stopper.

Injection of the local anesthetic to one anatomic area is followed by liposuction of the same area; that is, injection of one thigh is followed by its liposuction after which the procedure is repeated on the other thigh. The fat-bound lidocaine[174] is thus removed in a relatively short period and its toxicity is avoided. This permits the use of a much greater quantity of local anesthetic during a procedure.

The use of cryoanesthesia[88] and of neocryoanesthesia as recently described by Fournier[89] may diminish considerably the quantity of lidocaine required during the procedure. Fournier's method was modified so that the area to be treated is infiltrated with the local anesthetic diluted in saline solution chilled to 2C. The infiltration with this 2C solution not only has a cryogenic anesthetic effect but has a vasoconstrictor effect enhancing that of the Xylocaine and epinephrine as well. I discontinued the use of chilled plain saline (without Xylocaine) as the only anesthetic, as initially proposed by Fournier, because it is not sufficiently effective as the sole anesthetic or vasoconstrictor.

The anesthetic is administered by grasping and raising the skin and inserting a 20-gauge 3½-inch spinal needle in the subcutaneous fat (Fig. 10–4). The needle should be parallel to the skin and underlying muscle. The tip of the needle should not be so close to the skin that the local anesthetic could cause a wheal; this is not only painful but also creates irregularities in the skin, making the judgment of the final smoothness of the treated area difficult. Similarly, the injection of anesthetic into the muscle is equally painful and can cause injury to the underlying structures, such as nerves, blood vessels, muscle, and tendons. Intravascular injection of the anesthetic should be avoided by injecting while inserting or withdrawing the needle in the subcutaneous fat; that is, the needle should not be stationary while the solution is injected. To avoid unnecessary skin insertion, the injections should be performed in a radial fashion.

To achieve anesthesia as well as vasoconstriction in the same level of subcutaneous fat in which the cannula will be inserted, the surgeon must follow the depth of the needle with the same care used in following the depth of the cannula, using the fingers as a guide (Fig. 10–5).

When sufficient anesthetic has been injected in one area, the needle is withdrawn and reinserted in another point from which a different section can be reached. In this manner, the area to be treated is anesthetized, including the area beyond the lipodys-trophic mass. This will permit feathering during the procedure for the gradual diminution in the level of liposuction beyond the lipodystrophic area.

While the local anesthetic is being injected, the patient may require additional diazepam. This is given in increments of 2.5 mg IV with a total dose for the entire procedure usually not surpassing 15 mg IV. The local anesthesia is administered using a closed system, first presented by Glogau[177] and modified by the author in which the bottle containing the anesthetic solution is attached to a vented IV tubing.* This in turn is connected to a three-way stopcock and tubing.† A 60-ml syringe is attached to the end of the tubing. Thus, the solution is withdrawn from the bottle into the syringe and from there injected directly into the subcutaneous fat (Fig. 10–6). This avoids the use of open containers and possible contamination and makes administration of the local anesthesia a smoother and faster procedure. After some anesthesia is obtained with the 20-gauge needle, an 18-gauge spinal needle may be used. Its use will facilitate administration of the solution in a relatively short period, permitting the full analgesic effect of Sublimaze (which lasts approximately 1 hour) during the liposuction. Furthermore, the use of a small-gauge needle and a smaller syringe will prolong the administration of the chilled saline solution, and the effect desired by its low temperature will be lost.

In an area in which the adipose layer is thinner, such as the ankles, calves, and sometimes knees, one may wish to use a 20-gauge spinal needle only to prevent the linear accumulation of fluid. Table 10–1 indicates the quantity of local anesthetic required in different areas of the body.

---

*Venoset 78 with Cair clamp. Abott Hospitals, No. 1881.
†Three-way stopcock with 20-inch extension set. Dart Industries, Number 17-0032.

Figure 10–1. The waiting room should be comfortable and decorated in a soothing manner.

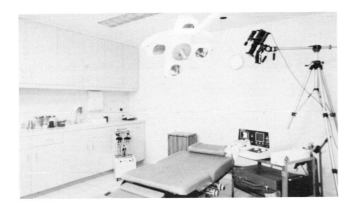

Figure 10–2. A partial view of one of the operating rooms at the author's surgical center.

Figure 10–3. A vial of fentanyl (Narcan) is kept taped to the IV pole.

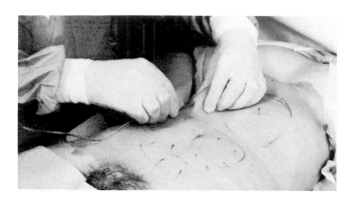

Figure 10–4. Local anesthesia is administered by grasping and raising the skin and inserting the spinal needle into the subcutaneous fat.

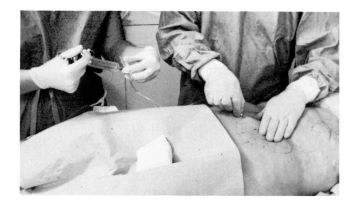

Figure 10–5. The depth of the needle can be felt with the hand not administering the anesthesia.

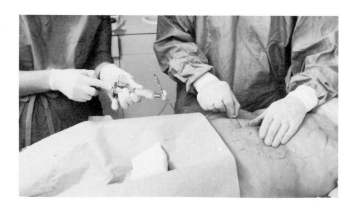

Figure 10–6. The closed system of local anesthesia administration uses a syringe with a three-way stopcock attached to the local anesthetic at one end and the spinal needle at the other end.

TABLE 10–1. QUANTITY OF LOCAL ANESTHETIC ADMINISTERED IN DIFFERENT AREAS OF THE BODY

Area	Quantity (ml)
Chin and neck	40–80
Arms	200–400
Flanks (love handles)	200–400
Hips	200–400
Thighs (saddlebags)	300–500
(liposuction to be performed on one thigh before commencing anesthesia on the contralateral thigh)	
Abdomen (upper or lower)	100–250
Buttocks	100–250
Knees	100–200
Calves/ankles	100–400

The maximum safe dose of lidocaine (with epinephrine) is approximately 7 mg/kg of body weight, although much greater quantities can be administered without untoward effects. Beyond this level, toxicity may occur manifested by convulsions or respiratory arrest.

Whenever possible, infiltration of one area should be followed by liposuction of the same area before proceeding to a different part of the body. The anesthesia can be administered in 10 to 15 minutes; when the entire area to be treated is infiltrated, one can return to the site in which the local anesthetic was first injected and can begin the liposuction. If large lipodystrophic areas are to be treated, such as thighs, only one area should be infiltrated at a time.

After the local anesthesia is administered and prior to beginning the surgical procedure, 1 ml of Sublimaze is again given in the manner described earlier. Thus, a total quantity of 2 ml of Sublimaze is given intravenously before surgery. This quantity of analgesic will suffice in most cases. If the patient becomes restless or experiences excessive discomfort, more Sublimaze is given in 0.5 to 1.0 ml increments. The maximum dose I have ever used was 6 ml (3 ampules, or 0.3 mg).

Fluid Replacement

11

Although the incisions caused by liposuction surgery are small, the procedure involves large areas. Because of the nature of the surgery, it can cause serious body fluid deficits. Constant monitoring of the patient, conservative approach, and full knowledge of the procedure will prevent complications. The presence of an anesthesiologist or anesthetist familiar with the medications used and fluid replacement therapy will ensure proper monitoring of the patient and early detection of any untoward reactions.

Liposuction involves the removal of fatty tissue with its natural milieu of serum, sometimes in great quantities. In addition, the destruction of fat cells during the repeated piston-like movement of the cannula within the adipose layer can be significant. Fournier and Otteni compared it with the crush syndrome.[118] The fluid loss is directly related to the degree of trauma; the quantity of fat destroyed is directly related to the quantity of fat removed. Thus, the fluid loss is related to the fat removed and the fluid administered during and after surgery represents replacement fluid rather than maintenance fluid.

It has been estimated that the average fasting patient will need 500 to 1000 ml of fluids before surgery to replace the insensible water loss that occurred during the previous 8 to 10 hours.[178] However, the technique described in this book does not require preoperative fasting.

Loading the patient with IV fluids prior to surgery is done by some surgeons to overhydrate and thereby cause hemodilution. By diluting the blood, fewer red blood cells are lost during surgery and the drop in postoperative hematocrit will be diminished.

Preloading the patient with IV fluids has not been found necessary when the technique outlined in this textbook is used. Nevertheless, the total quantity of tissue removed—consisting of fat, serum, injected anesthetic solution in saline, and some blood—must be replaced with intravenous fluids, usually lactated Ringer's solution and dextrose in water. The amount of replacement fluid I have found adequate during the past 5 years consists of the volume of tissue removed plus an additional 50 percent. Thus, if 1000 ml of "fat" is removed, the patient receives at least 1500 ml of replacement fluid.

The most common intravenous fluids given during a surgical procedure can be classified into two main groups: crystalloid and colloid. The crystalloid solutions include those containing various electrolytes as well as dextrose. The artificial colloid solutions, used as plasma substitutes, are those that contain various forms of dextrans (polysaccharides derived from glucose molecules) and Hespan (an intravascular volume expander derived from starch).

The main drawback of artificial colloid solution is the possibility of adverse reactions.[179–181] There are even more severe reactions that can potentially occur to albumin solutions, and many believe that they should be used only in the treatment of hypoproteinemia.[182,183]

In the average adult, 50 percent of body weight is formed by water. This is present both intracellularly (30 percent of body weight) and extracellularly (20 percent of body weight). The extracellular fluid is composed of 75 percent interstitial fluid (outside the cells or blood vessels) and 25 percent plasma present in the blood vessels. The interstitial fluid is considered functional because it participates in the maintenance of the homeostasis (Fig. 11–1).

Shock resulting from deficiency of functional fluids is caused by the fluids leaving the normal extracellular (interstitial and blood vessels) and intracellular location to be sequestered in a newly formed third space, now containing physiologically nonfunctional extracellular fluid. This is known to oc-

WATER = 50% of Body Weight

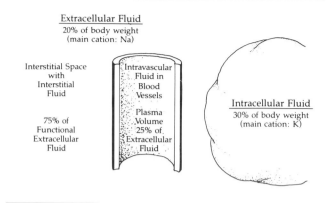

Extracellular Fluid
20% of body weight
(main cation: Na)

Interstitial Space
with
Interstitial
Fluid

Intravascular
Fluid in
Blood
Vessels

75% of
Functional
Extracellular
Fluid

Plasma
Volume
25% of
Extracellular
Fluid

Intracellular Fluid
30% of body weight
(main cation: K)

Figure 11–1. The schematic relationship between the extracellular fluid (containing the interstitial and the intravascular fluids) and the intracellular fluid is illustrated.

Shock due to Deficiency of Fluids

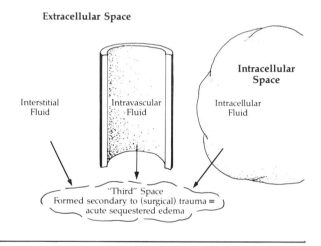

Extracellular Space

Intracellular
Space

Interstitial
Fluid

Intravascular
Fluid

Intracellular
Fluid

"Third" Space
Formed secondary to (surgical) trauma =
acute sequestered edema

Figure 11–2. Shock can result from fluid deficiency. Formation of a third space is attributable to (surgical) trauma, which withdraws functional interstitial fluid and plasma as well as intracellular fluid from normal metabolic functions.

Shock due to hemorrhage

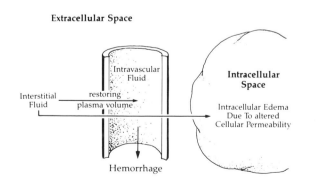

Extracellular Space

Intravascular
Fluid

Intracellular
Space

Interstitial
Fluid

restoring
plasma volume

Intracellular Edema
Due To altered
Cellular Permeability

Hemorrhage

Figure 11–3. Shock can be caused by hemorrhage. The diagrammatic changes taking place during hypovolemic hemorrhage shock are illustrated.

cur particularly in blunt trauma, such as the crush syndrome, as well as in trauma induced by surgery[184] (Fig. 11–2). This shock requires restoration of functional extracellular fluid, usually with a polyonic balanced salt solution (BSS), such as lactated Ringer's, also known as Hartman's solution. When these electrolyte solutions are administered in large quantities, however, as is necessary in liposuction surgery, peripheral edema may follow with the patient gaining weight after surgery.

The patient who suffers a severe loss of extracellular fluid may be apathetic, hypotensive, and tachycardic. The patient's stupor and hypotension may be wrongly attributed to the sedation received prior to surgery. Clinical signs also include dry mucous membranes and poor skin turgidity. Oliguria may be significant but not detectable when the surgery is performed under local anesthesia and a catheter is not used.

The gain in weight due to sequestered edema subsides in a few days with the use of diuretics. The fluid then passes from the third space into the functional interstitial space and from there into the renal circulation.

In shock caused by hemorrhage, the interstitial fluid passes from the interstitial space into the blood vessels to restore the plasma volume. Some of it, because of the altered permeability of cellular membrane, also passes into the cells, causing intracellular edema (Fig. 11–3). This type of hypovolemic shock requires several corrective measures:

1. Diminished extracellular interstitial fluid is corrected by treatment with a polyionic BSS (crystalloid therapy) such as lactated Ringer's solution. As much as 4000 ml may be required.

2. Plasma volume must be restored by the use of colloid therapy, such as whole blood or plasma expanders. The BSS often corrects the hypovolemia, however.

Thus, reestablishment of circulatory hemostasis with or without loss of blood requires restoration of both interstitial and intravascular fluids. This can usually be done with BSS, although plasma expanders and at times whole blood may be needed.[185] Hemorrhagic shock has also been treated with hyaluronidase, which corrects the interstitial tissue transport defect that occurs with acute loss of blood.[175,176]

Hemorrhagic shock is very unlikely to occur with the technique described here. Faulty technique and the use of traumatic instruments can increase the likelihood of states approaching shock caused by the loss of interstitial fluid into the third space, where it becomes nonfunctional and must be replaced.

The following fluid-replacement therapy is recommended:

a. Initial intravenous infusion consisting of 250 to 500 ml 5 percent dextrose in water as maintenance fluid therapy is given to replace the insensible loss and maintain the patient during the initial phase, that is, administration of local anesthesia, consisting of Xylocaine in NaCl.

b. Replacement fluid therapy is instituted, consisting of lactated Ringer's solution in a ratio of at least 1.5 : 1.0 of crystalloids to volume of aspirated tissue (fat, serum, blood, and local anesthetic injected in the adipose tissue).

Surgical Technique

12

Liposuction surgery involves the following steps:

1. The offending fat is removed.
2. The remaining fat is remodeled, especially peripherally, by feathering the surrounding area.
3. The skin is redraped by peripheral undermining, particularly in the abdominal area.
4. The skin is taped to reposition it, which helps create the new contour and avoid redundancy.

The goal of liposuction surgery is to contour the patient's face or body by removing unwanted deposits of fat. This should be done with a minimum of scarring and discomfort for the patient and a maximum of safety. To this end, local anesthesia and sedation before and during surgery have been found to be eminently effective in thousands of procedures (see Chapter 11).

This technique has been developed over a number of years. No one can claim credit for any particular procedure because each is a composite of the others that preceded it, and each surgeon brings about some degree of improvement.

One of the goals of cosmetic surgery is minimal scarring: liposuction is a surgical technique in which the scars can be small (4 to 6 mm in length) and hidden within a crease, fold, or another scar. If the incision is smaller than the diameter of the instrument used, the friction against the edges of the incision will injure the skin, and the resulting scars will be cosmetically not acceptable.

SURGICAL TECHNIQUE

After a 4- to 5-mm incision is made in a previously chosen and anesthetized area, blunt-end scissors are introduced to establish the plane in which the cannula will be inserted (Fig. 12–1). A 4-, 5-, or 6-mm cannula is then introduced, usually by grasping the skin, leaving an adequate layer of subcutaneous fat between the level of the cannula and the skin (Fig. 12–2). A cannula greater than 6 mm in diameter is rarely needed; a 4-mm cannula with three openings is usually adequate for most cases. The opening of this blunt-end cannula should always point downward, except in certain areas (see Chapter 14). The depth of the cannula level and the thickness of the superficial skin and fat will obviously depend on the area treated. For instance, the abdomen will require a heavier subcutaneous adipose layer than will the submental region.

Pretunneling is not necessary, and most experienced liposuction surgeons have abandoned its use. However, a 4-mm cannula (with the suction on) will facilitate penetration of the fat and the creation of tunnels. One can then change to a 6-mm instrument if the fat is rather abundant and finish with the 4-mm or smaller instrument for the creation of a finer network of tunnels.

After the instrument is introduced, the suction apparatus is started and the cannula is moved back and forth approximately 10 times in one tunnel. It is then partly withdrawn and reinserted radially, repeating the pistonlike movement. During this procedure, the left hand (assuming the physician is holding the cannula with the right hand) is held flat over the treated area, feeling and guiding the cannula (Fig. 12–3). Occasionally, the left hand will grasp the skin and underlying fat to facilitate the introduction of the cannula into a new level or a new direction. Often the left hand will exert countertraction on the skin to facilitate movement of the instrument within the fat (Fig. 12–4). In this manner, using the left hand as the artist and the right hand as the mechanic, liposuction is performed, molding, sculpting, and contouring the shape of the face and body. The use of smaller cannulae and the crisscross

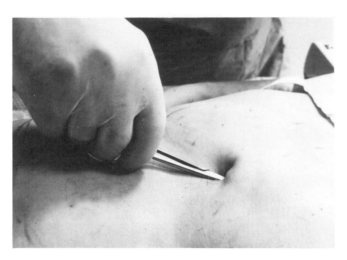

Figure 12–1. After the initial incision is made, a curved blunt-end scissors is introduced to establish the proper plane for liposuction in the subcutaneous fat.

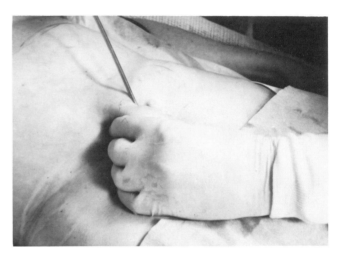

Figure 12–2. The cannula is inserted into the subcutaneous fat by grasping and raising the skin.

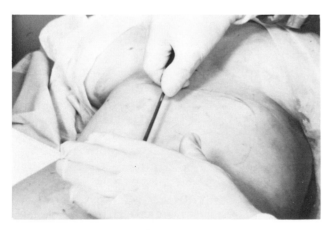

Figure 12–3. The guiding left hand held over the treated area should always feel the level of the cannula under the skin.

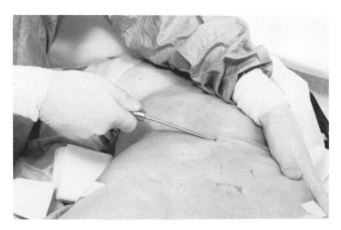

Figure 12–4. Countertraction with the left hand often will facilitate the movement of the cannula within the fat.

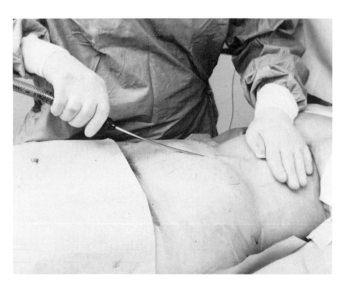

Figure 12–5. During the procedure, the surgeon should always know the depth of the cannula and, in particular, the location of its tip.

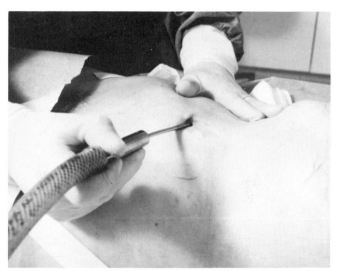

Figure 12–6. The cannula can be held lightly and maneuvered easily in liposuction body sculpting.

technique facilitates the creation of tunnels in several levels (swiss cheese or honeycomb effect), thereby avoiding surface irregularities such as indentations and valleys, which are more likely to occur when a larger-diameter instrument is used (Fig. 12–5). Throughout the procedure, the surface is constantly felt for irregularities (facilitated by wetting of the skin surface), and the skin is pinched to feel the thickness of the subcutaneous fat in different areas.

At the end of the procedure, kneading the skin and its subcutaneous fat will redistribute some of the loose fatty tissue left behind and may help achieve a smoother result. In effect, this is autologous fat transplantation without the adipose cells ever leaving the system.

The use of a sufficient quantity of local anesthetic injected within the adipose layer as well as the use of small-diameter instruments will make liposuction a creative experience rather than hard work. The cannula need not be forced into the fat, and it can be held as an artist's tool, like a brush or pen, to accomplish true body sculpting (Fig. 12–6).

During the surgery, whenever one cannula is exchanged for another, the one not used is placed in an 18-inch boat tray filled with cetylcide. During the procedure, the cannula should be wiped frequently with gauze soaked in cetylcide. Not only will this ensure sterility of the instrument, but it will also remove the fibrous tissue caught in the cannula apertures. The cannula can become clogged with fat or fibrous tissue, especially when an instrument smaller than 4 mm in diameter is used. A child's nasal syringe, pear shaped and made of rubber, can be used to push the air forcefully through the lumen and unclog the cannula.

When the cannula is detached from the connecting tube, it should be held horizontally (by the handle) to prevent the fat still present in it from spilling on the patient or the floor. Furthermore, when the nurse gives the instrument to the surgeon, it should be held only by the handle and not by the tube, to prevent possible contamination.

WHEN TO STOP

The decision of when to stop liposuction in one area can be made on the basis of the following factors.

1. *Achieving the desired contour.* One should not forget that a certain amount of adipose tissue that is destroyed by the mechanical action of the cannula is not aspirated. That tissue undergoes what Fournier calls a "biological resection." Liposuction that is too aggressive, although achieving what one may consider a pleasing contour at the end of the procedure, may result in defects that are difficult or impossible to correct.

   The surgeon performing liposuction surgery in effect sculpts the body into a new shape; it is truly body sculpting, and a good aesthetic sense is of paramount importance in achieving a good result. Often during the procedure physicians must step back and judge their work with the critical eye of a sculptor.

2. *Leaving sufficient fat under the skin for a pleasing aesthetic effect.* The thickness of subcutaneous fat can be judged by pinching the skin in various areas in which liposuction has been performed (Fig. 12–7). This action will ensure an even distribution of the adipose tissue and will also indicate to the operator when enough fat has been removed. It is obvious that different parts of the body require different thicknesses of the adipose layer. This is particularly important in an area in which the fat layer is not thick to start with, as is the case on the face, neck, or ankles.

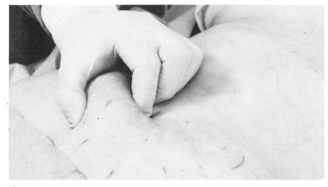

A

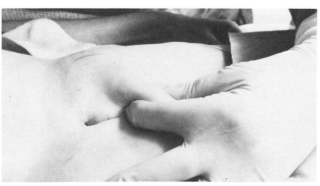

B

Figure 12–7. The pinch test of the skin and subcutaneous fat is performed both before and after liposuction.

3. *Quantity of blood present in the suctioned fat or the collecting bottle.* When the aspirated fat appears to contain more blood than is thought clinically safe, the operator should either change the area of surgery or the location of the cannula. The suction tube should be watched constantly to judge the quality of fat and the amount of blood being extracted (Fig. 12–8). The suctioned fat should not contain more than 15 percent blood. Suctioned tissue consisting of more than one third blood[186] is totally unacceptable regardless of where the procedure is performed and particularly when performed in an outpatient surgical facility. This would mean that for 2000 ml of fat removed the patient would lose 700 ml of blood. Such a case would obviously require immediate blood replacement. Figure 12–9 shows the quality of fat that should be obtained with liposuction performed under local anesthesia.

CLOSURE

The incisions are usually so small that three or four sutures, 5-0 or 6-0 nylon, are sufficient to close the wounds. Buried absorbable sutures have not been found necessary because the length of the incision rarely goes beyond 8 mm in length.

I have never found the use of drains necessary; most surgeons experienced in liposuction surgery technique do not use them. There may be some drainage at the site of the incision, and an extra padding of gauze should be placed there. Patients should be told that a certain amount of drainage is normal and that they will have to change the gauze once or twice.

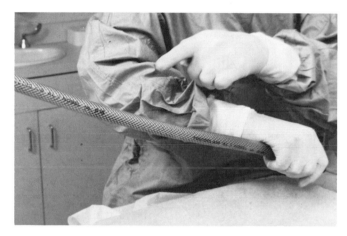

Figure 12–8. The fat in the suction tube should be watched constantly for excessive blood.

A

B

Figure 12–9. The quality of fat that should be obtained by liposuction performed under local anesthesia is shown.

Possible Complications

13

INTRAOPERATIVE COMPLICATIONS

Complications Caused by Medication

It seems redundant to mention that a physician using a particular medication should be thoroughly familiar with its pharmacologic action and side effects. In addition to this knowledge, the physician should have all the necessary medication on hand so that, if necessary, remedial action can be taken swiftly and effectively. Resuscitative apparatus should be available as well, and the physician should know how to use it. Basic knowledge of cardiopulmonary resuscitation (CPR) is essential for all personnel, and the surgeon should make all possible efforts to complete a course on advanced life support.

Setting the table in the Trendelenburg position and speeding the flow of the intravenous infusion will usually be enough to counteract a vagovagal reaction or hypotensive episode. Although anxiety or a chilly room may cause the patient to shiver, the same effect may result if the local anesthetic reaches the toxic level.

Diazepam may cause phlebitis and venous thrombosis, so its injection should be slow. Administration of the intravenous medication using a tuberculin syringe with a small-gauge needle will help ensure its slow injection. Its dose should be diminished accordingly, depending on the age of the patient and on concomitant use of other medications, particularly narcotics.[138,139]

Fentanyl may cause muscle rigidity, particularly of the chest. This complication may be related to the speed of injection; the technique described above should be employed for its administration.[160-162]

Any surgical procedure, particularly when done in an outpatient setting, should be undertaken with utmost care and prudence and with full knowledge of all possible side effects and appropriate counteractive measures. The employment of a highly trained person, such as an anesthesiologist or anesthetist, is strongly suggested to any physician who contemplates performing liposuction in his facility. Standard textbooks should be consulted and reviewed by the surgeons employing agents used in this procedure.

Complications Caused by the Procedure

Aside from the removal of too much fat by liposuction (Fig. 13–1), blood vessels and nerves can be injured.

Perforation of the abdominal musculature with consequent injury of the underlying structures is always possible; therefore, the tip of the cannula should always be palpable or visible (Fig. 13–2). Bowel perforation can occur even when proper technique is employed, particularly when there is significant diastasis of recti muscle which may be present with or without a ventral bermia. The latter may become apparent only after theepigastric fat, previously masking it, is removed.

The surgeon should always be thoroughly familiar with the anatomy of the area in which liposuction is performed and should be aware of aberrant anatomy (Fig. 13–3). Facial liposuction can be particularly perilous in this regard.[187]

POSTOPERATIVE COMPLICATIONS

Fatalities have been reporte due to necrotizing fascitis, severe infection, hypovolemic shock, disseminated intravascular coagulopathy, pulmonary emboli, fat emboli and fat embolic syndrome.[187a]

Pulmonary (blood) emboli can be a significant threat particularly when liposuction is combined with other procedures such as abdominoplasty. In addition, prolonged immobilization and below-the-knee constrictive dressings may enhance this possibility, which may have been the case in a 62-year-old patient of a plastic surgeon. Liposuction performed under local anesthesia allows immediate postoperative mobilization which is not usually possible when general anesthesia is used.

Fat emboli can also occur; as previously mentioned, Fournier and Otteni compared the liposuction procedure with the crush syndrome.[118] It seems obvious that the use of a large and nonaerodynamic blunt-tipped instrument will increase the destruction of the fat cells, which may facilitate the formation of fat emboli. As stated, fat embolism is significantly more likely to occur when liposuction is performed in conjunction with an abdominoplasty.

Fat embolic syndrome is rare and usually occurs within 24 hours. The classic pettechiae on the anterior axillary folds and flanks are due to thrombocytopenia caused by adherence of platelets to circulating fat globules. Furthermore, the fat emboli trapped in the lungs are hydrolyzed by the pneumatocytes' lipase into glycerol and free fatty acids (FFA). The latter change the permeability of the pulmonary vasculature allowing the flow of high-protein fluid into the interstitial and alveolar spaces causing pulmonary edema and associated manifestations. Chest x-rays, arterial blood-gas measurements, calcium level monitoring (which has an affinity for FFA), and immediate steroid treatment are essential.

The postoperative complications most commonly reported are gross irregularities in the skin surface, contour deformities, hyperpigmentation, hematomas and seromas, and loose skin. Edema and bruising should not be considered complications unless they are excessive and long-lasting. Betamethasone sodium phosphate, 6 to 12 mg IM, before or after surgery will help prevent or diminish postoperative edema. Similarly, massage and ultrasound therapy two to three times per week following surgery will help resolve the edema, ecchymoses, and induration within a shorter period. These procedures will also make the patient feel more comfortable during the postoperative period.[188–190]

Irregularities in the Skin Surface

Assuming the patient does not have cellulite, postsurgical irregularities in the skin surface, such as waviness, appear mainly on the thighs, both laterally and medially. This problem is usually caused by insufficient turgor and contractability in the patient's skin to counteract the surgeon's overzealousness. In addition, the consistency of the fat (too "soft" and exhibiting ballottement) contribute to these poor results. Fischer was the first to suggest that vertical tunneling may help prevent the cascade of horizontal waves that can occur following liposuction. However, if these waves are present before surgery, it is hardly likely that liposuction will remove them.

Gross irregularities are caused by excessive removal of fat and insufficient consideration of the biologic resection of fat left behind while the surgery is being performed. Whereas excessive removal of fat is visible immediately, absorption of the injured fat tissue may take weeks or months to complete. Consequently, the changes occurring because of the resorption of injured adipocytes may not be obvious until several months after liposuction surgery.

Fischer advocates collecting aspirated fat in sterile containers. If a defect induced by surgery is noted immediately, it can be corrected by reinjection of the removed fat (see Chapter 16).

Irregularities in the skin surface caused by the nodular consistency of fat can be especially apparent in the epigastric area, where the adipose tissue is more fibrous than in other areas. Occasionally, a particularly indurated area may respond to triamcinolone acetonide, 10 mg/ml, injected in the hard or protuberant tissue.

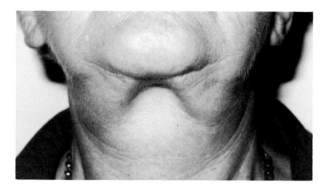

Figure 13–1. Too much fat was removed from this patient's submental area by a "qualified," board-certified plastic surgeon, resulting in a serious deformity.

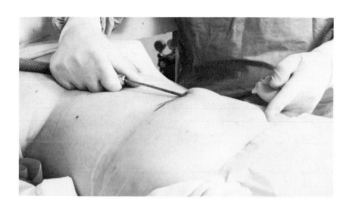

Figure 13–2. The surgeon should always know the location of the tip of the cannula. In addition, a cannula with a blunt tip is less likely to cause depressions under the skin even when its tip reaches the dermis.

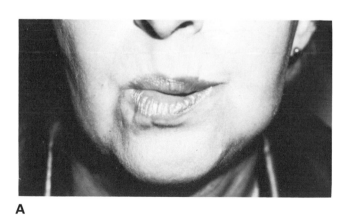

A

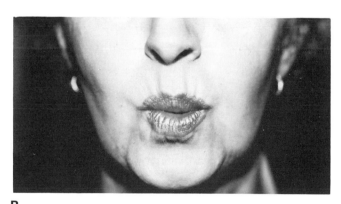

B

Figure 13–3. Temporary paresis was caused by bruising of an aberrant branch of the submandibular nerve. The surgeon should be thoroughly familiar with the anatomy of the area being operated on.

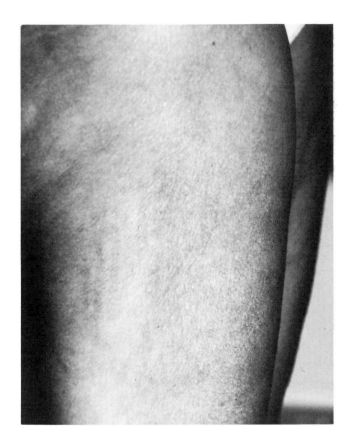

Figure 13–4. Striated hyperpigmentation, although mild, can occur following liposuction of the thighs.

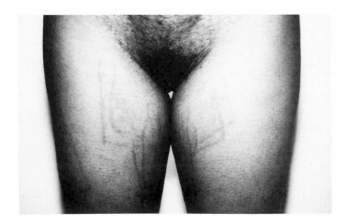

Figure 13–5. In this dark-skinned patient, postinflammatory hyperpigmentation resulted following taping.

Contour Deformities

Resorption of injured fatty tissue may take months to complete. Contour deformities are mainly attributable to this occurrence. Results of the procedure may appear excellent, yet months afterward these contour deformities appear, particularly in the posterolateral thighs (actually lateral buttocks), the gluteal fold, giving the appearance of "dropped" buttocks, and lower legs, which may appear irregular and do not have a pleasing shape.

Hyperpigmentation

Hyperpigmentation of the skin over or around the area operated on may occur for two reasons.

1. Long-lasting ecchymoses will cause deposition of hemosiderin in the above areas, particularly in the vicinity distal (dependent) to the operated site. This hyperpigmentation can be either diffuse or striated, following the paths of the cannula (Fig. 13–4). Early massage and ultrasound therapy shorten the duration of ecchymoses and consequently diminish the likelihood of this problem.
2. Application of tape on hyperpigmented skin may cause postinflammatory hyperpigmentation, which may be long-lasting and difficult to diminish (Fig. 13–5). The use of a garment without tape, Kurlex (Kling) bandages (especially around thighs) and application of bleaching agents may be beneficial.

Hematomas and Seromas

Hematomas can occur during liposuction surgery, although I have not personally seen them in any of my patients. Several precautions are recommended:

1. Use epinephrine to diminish the possibility of bleeding.
2. Avoid trauma to the large vasculature by being thoroughly familiar with the anatomy of the area and staying within the right plane.
3. "Milk" the area operated upon of any accumulated blood before suturing the incision and applying the dressing.

Seromas were rather common when liposuction was performed with an aspirating curette. Seromas are localized accumulations of serous fluid caused by traumatized lymphatic vessels. Their traditional treatment includes frequent aspirations or the use of drainage catheters. The latter increases the risk of infection. More recently, seromas have been treated with a tetracycline solution (2 gm/150 ml saline) injected in the space left by aspirating the serous fluid. The solution was removed after 45 minutes, after which closed suction drainage and pressure dressing was applied.[191] Apparently this type of sclerotherapy has not been used for the treatment of postliposuction seromas, although it has been used for pleural effusions.[192,193] Since the advent of blunt-tip cannulae in liposuction surgery, the frequency of seromas has diminished dramatically. The careful application of tape or a tight garment can also prevent their occurrence.

Loose Skin

Loose skin can occur when liposuction is performed on a patient who already has excessive or inelastic skin. It can also occur when the cannula is used in a lateral rather than back-and-forth movement. This windshield-wiper movement creates a large space between the skin and the underlying tissue. The serous exudate of the damaged fat can prevent the adhesion of the large flap of skin to the underlying tissue, particularly when the postoperative dressing is not applied properly.

Suggestions for Liposuction of Individual Areas

14

There is no original liposuction technique. All approaches are composites of earlier techniques modified and improved to fit the surgeon's individual temperament and surgical philosophy.

There are probably as many approaches to liposuction surgery as there are surgeons performing it. However, I have found the following techniques most consistently successful.

FACE AND NECK

A submental incision for liposuction of the neck can accentuate the vertical bands of skin under the chin; excessive suction of submental fat also can create a true deformity by causing adhesion of the skin to underlying muscle.[187] Other surgeons, however, have found the submental incision quite adequate[194] and, if properly performed, liposuction using this incision can give excellent results. It can also be used to reach the jowls if the surgeon chooses not to use other incisions (Fig. 14–1).

Two small incisions, each 3 to 4 mm, can also be made on either side of the neck, preferably within a crease (see Fig. 9–4). Alternatively, they can be made in the preauricular area, in front of the lobule or just under it. There, a natural crease can always hide the incision. The preauricular incision is indicated to reach the fat in the jowls and cheeks (Fig. 14–2). One should be wary of possible injury, usually temporary, to the submandibular branch of the facial nerve (see Fig. 13–3). From the preauricular area, one can reach the submental area and beyond it (Fig. 14–3). When this is done from both sides, a crisscross pattern of tunneling is created under the chin, thus avoiding the above-mentioned submental vertical folds that can follow the submental incision alone (see Fig. 9–4). If a submental incision is made, care must be taken against excessive suction in the midline, which can cause a serious deformity in this area. Figure 13–1 shows the result obtained by the too-aggressive technique of a "qualified" plastic surgeon.

I do not follow the technique of complete separation of skin under the chin, which in effect follows Kesselring's technique of undermining a large area with continuous suctioned lipectomy; it may increase the chance of seromas and hematomas.

When removing excessive fat from the neck and submandibular area, the operator may wish to turn the cannula with the opening toward the skin to thin the subcutaneous fat sufficiently to give the patient a pleasing contour of the jaw line. Figure 14–4 shows pre- and postoperative views of four patients who underwent liposuction of chin, jowls, and neck. This technique should be employed with utmost care because the adhesion of fat-free skin to the mandible can be disfiguring and can give a skeletonized effect.

Figure 14–5 demonstrates before and after views of patients who underwent the combined procedures of liposuction of chin and neck and a chin implant, usually placed through an intraoral route.

The fat in the melolabial fold can be easily reached through a 2 to 3-mm cutaneous incision made directly in the fold behind the nostril, or within the nasal antrum in the pyriform recess, which will heal in 2 to 3 days without sutures.[195,196] A 2- or 3-mm cannula can be used for this purpose. One must warn the patient before surgery that because of the thickness of the skin just lateral to it, the melolabial crease may remain evident, even though the fat lateral to it is removed. Similarly, although the malar fat pads can be easily removed through temporal incisions, the edema and postinflammatory hyperpigmentation can last for months.

Liposuction can also be used as an adjunct procedure during a rhytidectomy. This was first reported in 1982 by Plot at a meeting of The French Society of Esthetic Surgery.[197] Subsequent articles by both French and American surgeons confirmed the value of this procedure.[117,198–200]

Liposuction surgery or technique can be used during rhytidectomy in several ways:

1. Liposuction of the neck and chin prior to the routine rhytidectomy obviates the need for the traditional submental lipectomy with its large incision.
2. A routine rhytidectomy can be performed and the fat present on the parotid fascia and superficial musculo-aponeurotic system (SMAS) prior to its plication or imbrication can be "cleaned up" using a flat cannula with its opening against the parotid fascia. With the suction on, the cannula is rubbed gently against the SMAS and fascia, thereby suctioning off the excess as well as the residual fat left on them. The flat cannula may have certain advantages over the traditional round one in this procedure.[201]

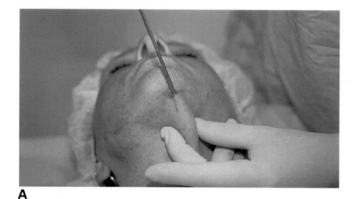

A

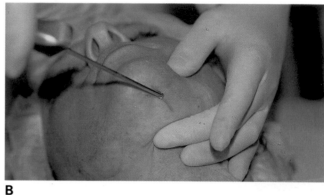

B

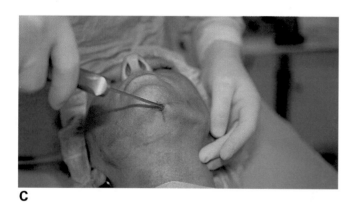

C

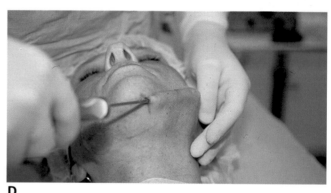

D

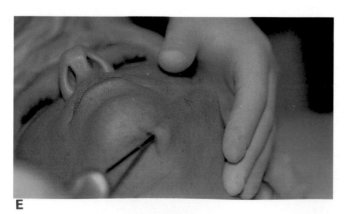

E

Figure 14–1. The submental incision can be used for liposuction of chin and jowls. The cannula must be kept away from the mandibular ramus.

3. Using the liposuction technique, but not the procedure itself, the surgeon uses the cannula, either round or flat, for blunt dissection of the face and neck and to raise the flap prior to the routine continuation of the rhytidectomy. One should follow the good technique of keeping the instrument tip against the skin to avoid damaging underlying structure such as nerves and blood vessels.

Any combination of these approaches can be used, depending on the surgeon's experience and the patient's condition. I have found it particularly helpful to use the blunt dissection method (using a flat cannula) to separate the congenitally nondecussated platysmal muscles[202] from the overlying skin. These are present at the junction of the posterior aspect of the undersurface of the chin and the upper anterior aspect of the neck (Fig. 14–6; see also Fig. 5–7). Quite often, the rhytidectomy result is enhanced by a chin implant, if indicated (Fig. 14–7).

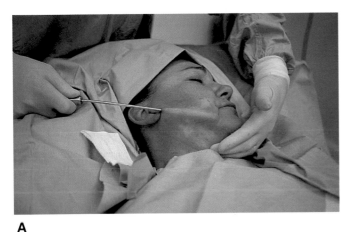

A

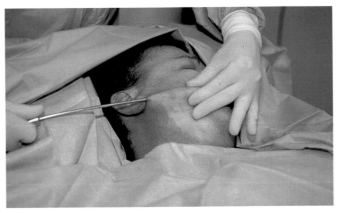

B

Figure 14–2. Liposuction of jowls and cheeks can be performed through a preauricular incision. The cannula is kept away from the deeper structures in those areas to prevent injury to the nerves.

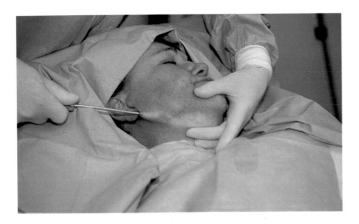

Figure 14–3. The preauricular incisions can be used for liposuction of the chin and neck, creating a crisscross pattern of tunneling in the undersurface of the chin (see also Fig. 9–4).

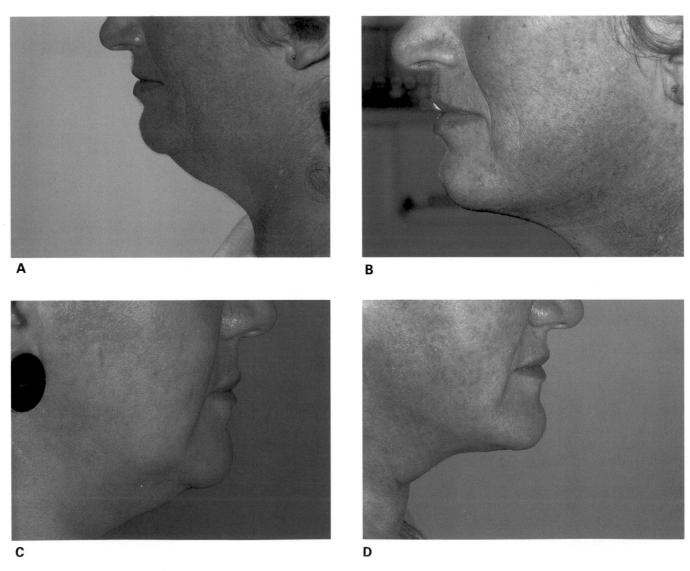

A

B

C

D

Figure 14–4. The results of liposuction of chin, jowls, and neck in four patients, shown in pre- and postoperative photos.

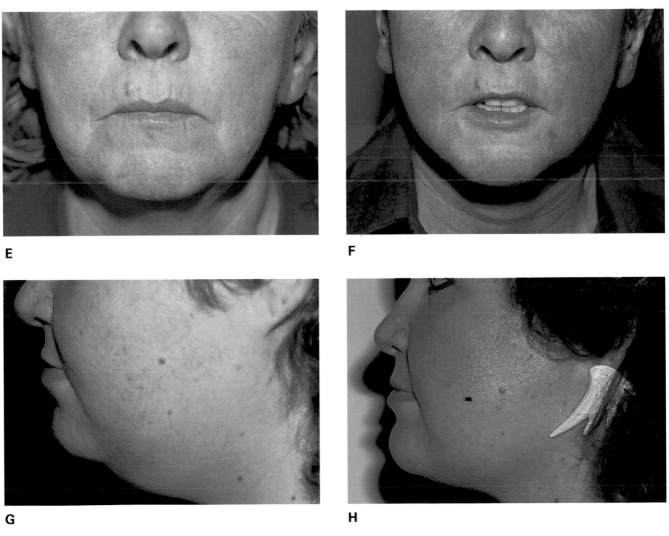

Figure 14–4. *Continued.*

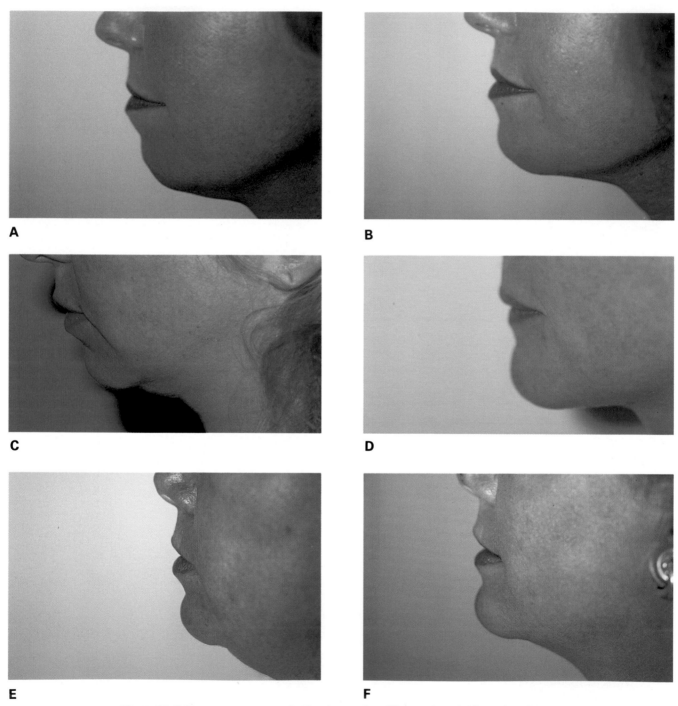

Figure 14–5. To correct a recessed chin, the results of liposuction of chin and neck can be enhanced using a chin implant.

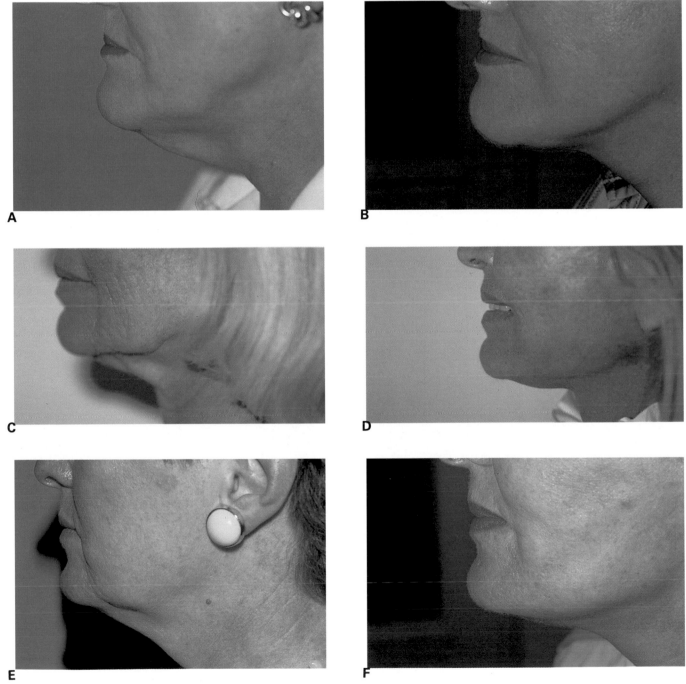

Figure 14–6. Pre- and postoperative photographs show patients who underwent rhytidectomies with the help of either one, two, or all three techniques described in the text.

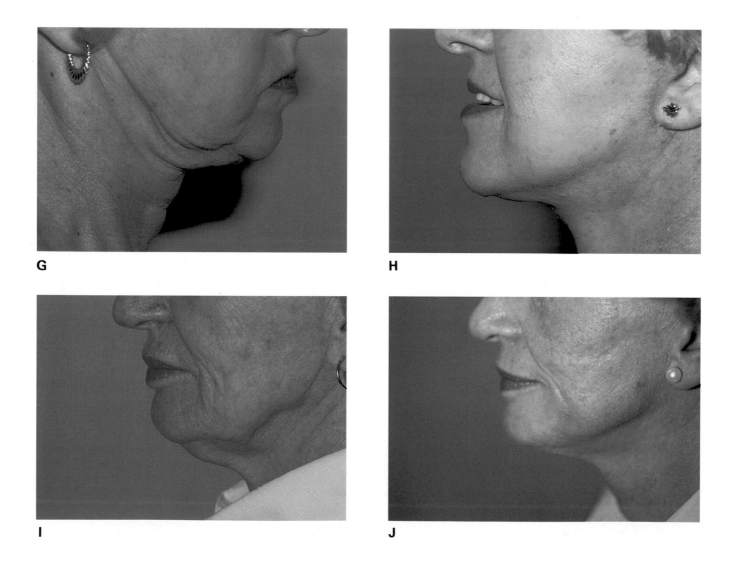

G

H

I

J

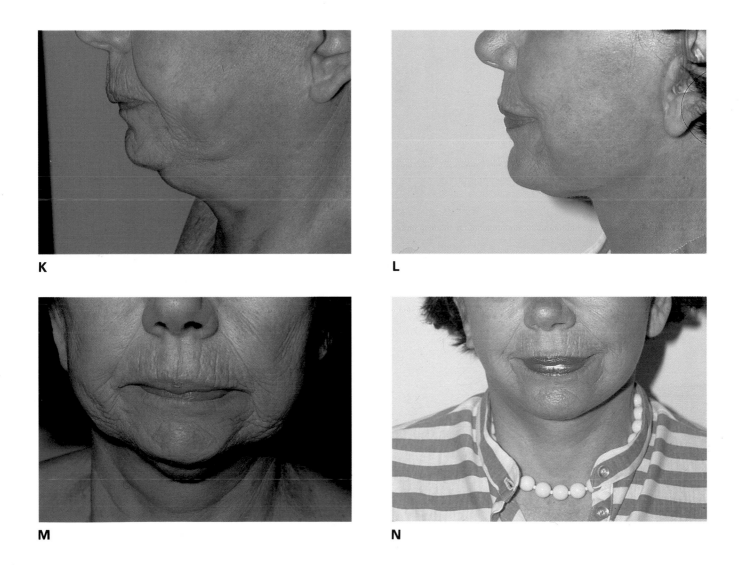

K

L

M

N

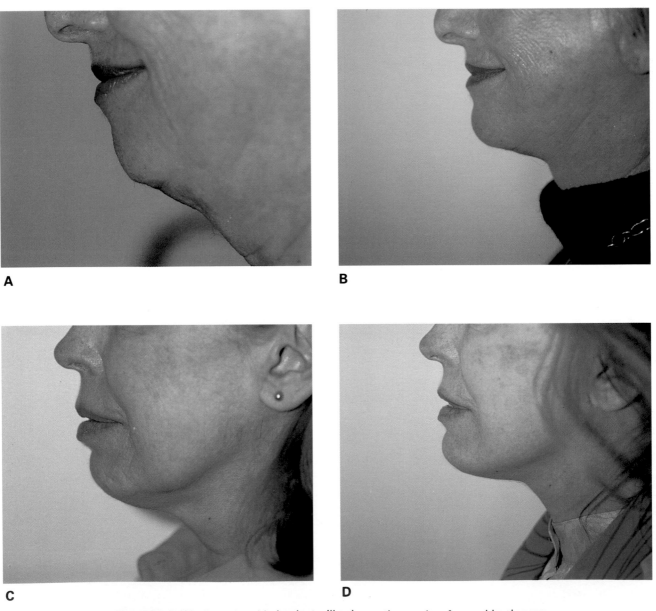

Figure 14–7. If indicated, a chin implant will enhance the results of a combined procedure of liposuction and rhytidectomy.

ARMS AND MEDIAL THIGHS

Liposuction of the arms is facilitated when the incision is made above the elbow rather than in the posterior axillary fold, although occasionally both incisions are necessary (Figs. 14–8 to 14–10). The inner thighs can be approached from the medial aspect of the gluteal fold, as well as from the femoral fold, although there the protuberance of the abdomen may interfere with the free movement of the hand unless the operator uses a curved or a Paragon* cannula. Liposuction of arms and medial thighs should be deferred until the operator has sufficient experience with this procedure. They are both nonforgiving areas.

Occasionally, fatty tissue present in the upper medial thigh can be better reached through an incision in the medial aspect of mid-thigh. The location of the incisions should be thoroughly discussed with the patient before surgery is performed.

---

*Robbins Instruments, Chatham, N. J.

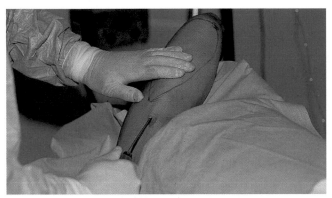

Figure 14–8. The posterior axillary fold approach can be used for liposuction of the arm.

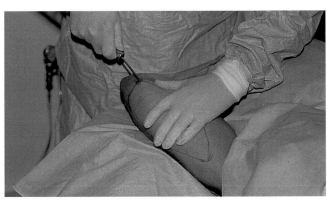

Figure 14–9. An additional incision in the posteromedial aspect of the elbow will facilitate liposuction of arms and will enable the surgeon to perform crisscross tunneling.

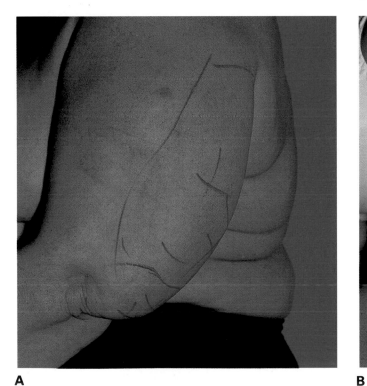

A

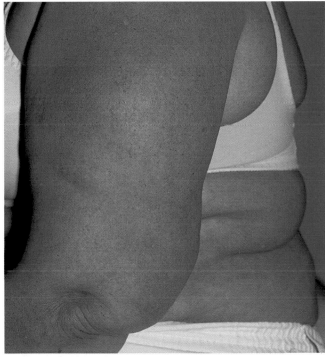

B

Figure 14–10. Pre- and postoperative views of an arm that underwent liposuction are shown. As excessively heavy arms made it impossible for this patient to wear a blouse, surgery was performed for functional as well as cosmetic reasons.

CHEST

In men the liposuction of the chest, either with true or pseudogynecomastia, may be accompanied by the surgical excision of glandular tissue, which can be reached through a semicrescentric incision in the lower aspect of the areola. The Cobra* cannula, however, has changed the approach to this area. Because of its structure, it can penetrate, break up, and aspirate the adipose as well as the glandular tissue. A 4- to 5-mm incision is made in the upper aspect of the anterior axillary fold through which the cannula is inserted into the excessive adipose and glandular tissue (Figs. 14–11, 14–12). Liposuction is started with a 4-mm cannula, which can penetrate the mammary tissue more easily than the 6-mm cannula.

After sufficient tunnels have been made (with the suction on), the procedure is continued with a 6-mm Cobra cannula. From time to time, the surgeon may want to use a three-aperture cannula. Holding it with its openings turned upward toward the skin and using it as a rasp (according to Fournier's concept), the surgeon may be able to remove the rest of the glandular tissue attached to the areola in the form of a disk (Figs. 14–13 to 14–15). Before surgery, the protuberant lateral mass of the pectoral muscle should be pointed out to the patient to avoid disappointment afterward for not having a completely flat chest (see Figs. 7–5, 7–6).

---

*Wells-Johnson Co., Tuscon, Ariz.

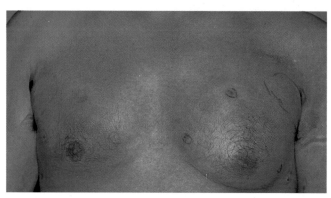

Figure 14–11. An intraoperative photograph of a patient undergoing a subcutaneous liposuction mastectomy for pseudo gynecomastia is shown. Markings indicate where local anesthesia of the skin was performed. Anesthesia of subcutaneous fat and glandular tissue is performed through the marked areas.

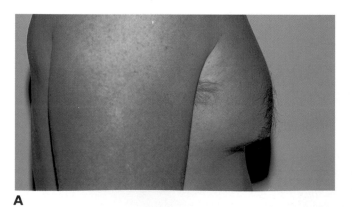

A

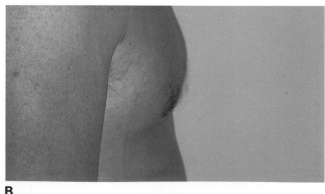

B

Figure 14–12. Before and after views are shown of the same patient in Fig. 14–11.

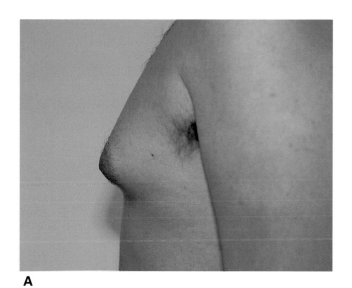

A

B

Figure 14–13. True gynecomastia in a 28-year-old man was corrected with liposuction through an anterior-superior axillary incision.

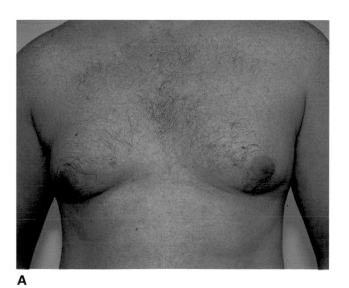

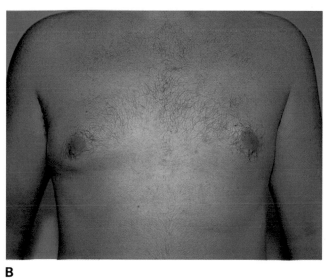

A

B

Figure 14–14. Pseudo gynecomastia in a 32-year-old man was corrected with liposuction surgery using the anterior axillary line approach.

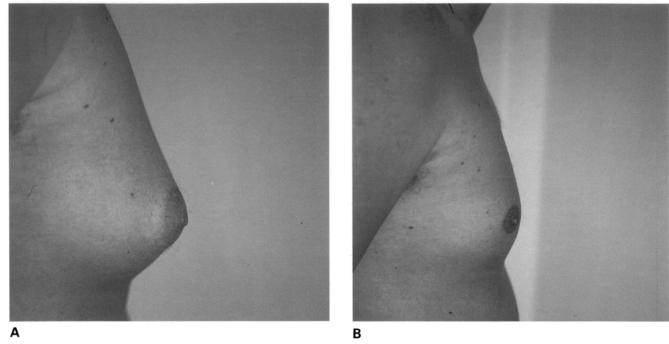

A   **B**

Figure 14–15. True gynecomastia in a 23-year-old man was corrected with liposuction.

UPPER ABDOMEN

The epigastric area, particularly in men, has more fibrous tissue than the infraumbilical area, and the operator will encounter greater resistance to the cannula. This is also an area in which more bleeding may take place. The upper abdomen can be reached either through an umbilical incision or through two suprapubic incisions with a long cannula. Because of the elevation of the lower part of the rib cage, particularly in women, the dressing requires special attention (see Chapter 15).

LOWER ABDOMEN

The protuberance of the abdomen may be partly attributable to the weakness of its musculature. Most of the abdominal fat can be easily reached from an umbilical incision; however, the periumbilical area may be more easily reached and its fat removed through one or more incisions made in the suprapubic region (Figs. 14–16 to 14–20).

When performing liposuction on a large abdominal panniculus, usually in stages, several incisions may be necessary in the crease of the apron or just below. There, above all, the tunneling should be vertical and perpendicular to the hanging fold to prevent postoperative waves and to increase the chance of proper contraction of the skin with its subcutaneous fat (see Figs. 5–5, 5–6, 7–10, 7–11). Tunneling should be done from the deeper layers toward the surface. In staged liposuction, during the first stage of surgery, liposuction should be performed in the deeper layers of the fat. The second stage should be addressed to the more superficial layers, and so on. Contraction of the tissue will then take place from the deeper layers toward the surface and will avoid the formation of loose and pendulous skin as well as irregularity of the abdominal surface (Fig. 14–21).

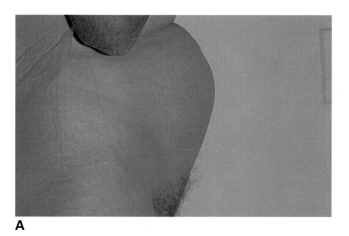
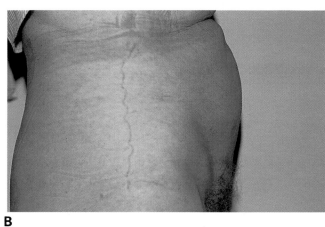

A **B**

Figure 14–16. Liposuction in a 68-year-old woman can give satisfactory results if the choice of patient is made judiciously.

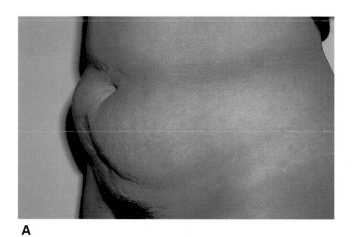
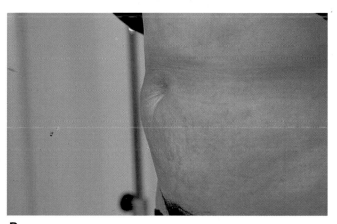

A **B**

Figure 14–17. Compare these before and after liposuction photographs of a patient who was previously told that an abdominoplasty was the only procedure that could improve her appearance.

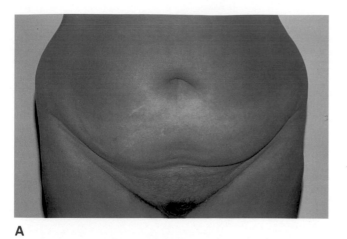
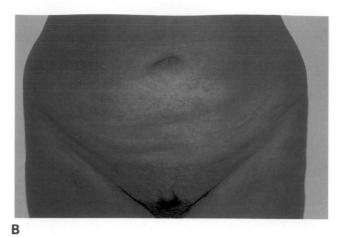

A

B

Figure 14–18. Results obtained in a pendulous abdomen with careful liposuction using a small-diameter cannula (one-stage operation) are shown.

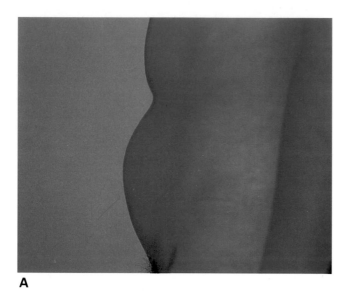
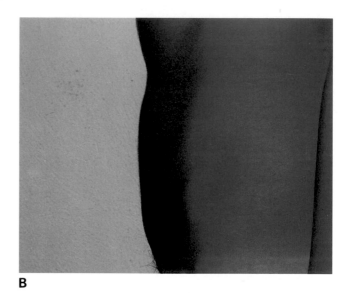

A

B

Figure 14–19. Liposuction of the abdominal wall was performed in this 27-year-old woman.

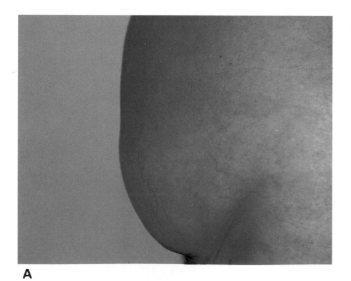
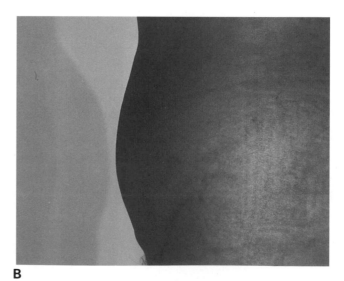

A

B

Figure 14–20. Liposuction in a patient with weak abdominal musculature can greatly improve the contour, even though the abdomen is not completely flat. Judicious choice of patients is absolutely essential.

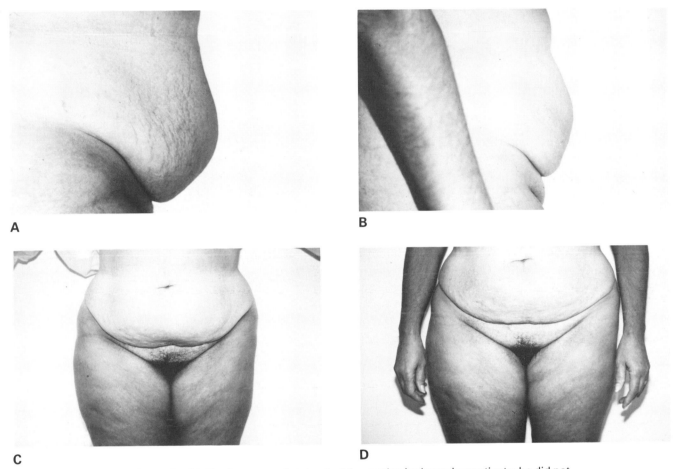

Figure 14–21. The first stage of abdominal liposuction is shown in a patient who did not want an abdominoplasty. A period of 3 to 6 months between interventions is advisable.

THIGHS AND HIPS

Excessive trochanteric fat, the so-called saddlebags or riding breeches, is the most common complaint of women. As pointed out so well by Fournier, simple removal of excess fat from the thighs will not improve the contour to its best. The adipose tissue present in the lateral lower buttocks should also be removed to prevent its lateral weight from bringing the thighs down again (see Figs. 9–10, 9–12).

The Pitanguy procedure, which employs a fusiform resection of skin and fat involving the inferolateral aspect of the buttock and the gluteal fold, has been used for many years to achieve a buttock lift and to improve the saddlebag appearance.[203] Fournier ingeniously employs the removal of fat with liposuction alone from the same area to achieve the buttock lift previously possible only with Pitanguy's operation. This cuneiform removal of fat from the buttocks, gluteal fold, and trochanteric area with liposuction alone is called by Fournier and Asurey the closed Pitanguy[204] (Figs. 14–22, 14–23).

For this procedure, I prefer to have the patient in the prone position; supratrochanteric incisions will permit the hips and flanks as well as the thighs to be reached. In addition, a small incision may be necessary in the lateral aspect of the gluteal fold to reach the posterior thigh as well as the buttock; this will also permit a crisscross pattern of tunnelization. If the crease of the buttock is too short, it can be extended laterally and superiorly by turning the cannula so that its opening is upward toward the skin and using it as a rasp against the dermis to create the fold and produce the desired adhesion between the skin and subcutaneous fat (Fig. 14–24). Again, utmost care must be exercised when the cannula is turned around.

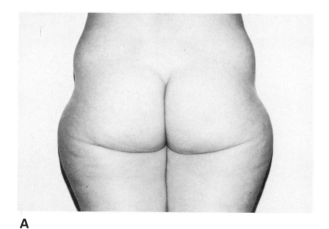

A

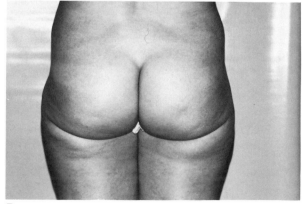

B

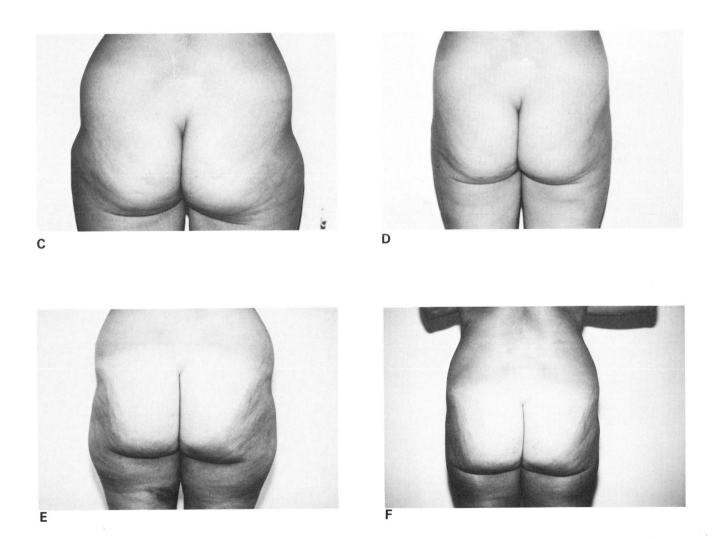

Figure 14–22. Pre- and postoperative views of patients who underwent liposuction for correction of trochanteric lipodystrophy are shown. The closed Pitanguy procedure (removal of adipose tissue from lateral lower buttocks) was performed in all cases shown.

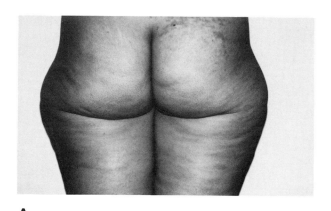

A

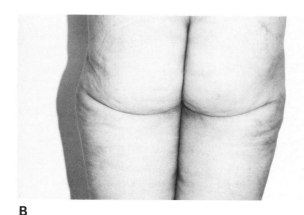

B

Figure 14–23. The persistence of cellulite is demonstrated after liposuction of the thighs, although the contour was dramatically changed. The patient was warned of this and was grateful with the result obtained.

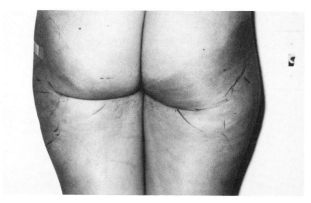

Figure 14–24. The gluteal fold can be accentuated, lengthened, or created with careful liposuction and by turning the opening of the cannula toward the skin. Compare the right side (preoperative) with the left side (postoperative).

Occasionally, liposuction of thighs and hips is performed to correct deformities caused by a traditional thigh lift (see Figs. 5–8, 5–9).

FLANKS

Excessive fat in the so-called love handles is one of the most common complaints by men (Figs. 14–25, 14–26). It should be pointed out to the patient that the anterior aspect of this deposit is really the iliac crest, which usually contains more skin and fibrous tissue than fat. Local anesthesia in this area should be started from the posterior aspect because of the course of the intercostal nerves. Here, too, the fat is rather fibrous and one can encounter resistance to the cannula as well as bleeding.

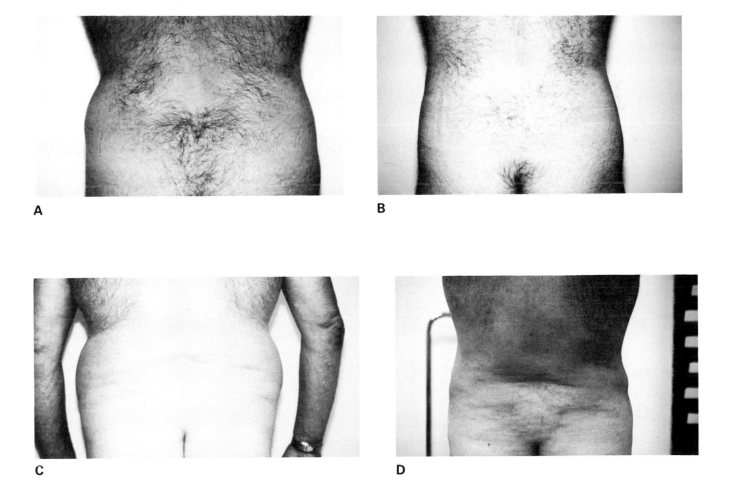

Figure 14–25. Adipose tissue in a man's flanks can be moderate to excessive. Liposuction in this area can be most successful (see also Figs. 7–13, 7–14).

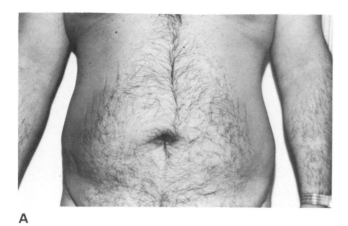
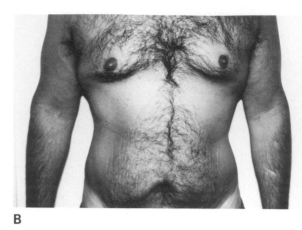

A   **B**

Figure 14–26. Compare these pre- and postoperative photographs of a patient who underwent staged liposuction several times for correction of lipodystrophy.

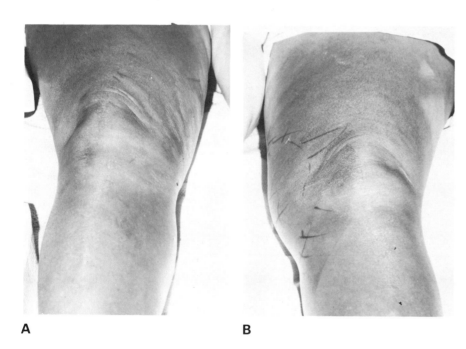

A   **B**

Figure 14–27. The left knee (postoperative) is shown, contrasted with the right knee (preoperative), demonstrating the change brought about by liposuction.

KNEES, CALVES, AND ANKLES

Knees are probably the most dramatic areas on which to perform liposuction because the patient can immediately see the difference between the knee that has been operated on and the one that has not (Fig. 14–27). Incisions in this area should be slightly posterior to the medical aspects of both knees and within a crease (see Fig. 9–13). Massive lipodystrophy of the knees can also be corrected with liposuction (Fig. 14–28).

There is one area on the medial knee in which the surgeon should be particularly careful in performing liposuction. After examining the knee with the leg extended, the surgeon should watch for the appearance of a naturally occurring depression upon bending the knee. This declivity is present medial and just posterior to the medial epicondyle of the femur. This depression in the skin should be marked in a different-color pen, so that the surgeon can see the movement and the overlapping of the fatty deposit in the medial knee over this valley.

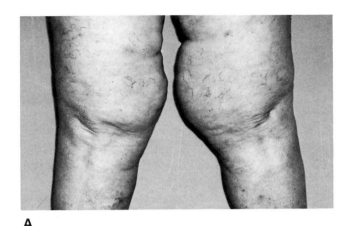
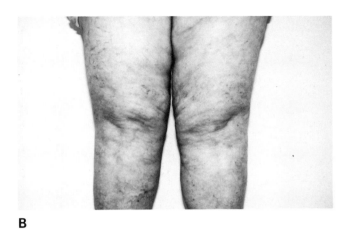

A

B

Figure 14–28. Massive lipodystrophy of the knees was corrected with liposuction. This patient required staged liposuction (two procedures) to obtain the result shown.

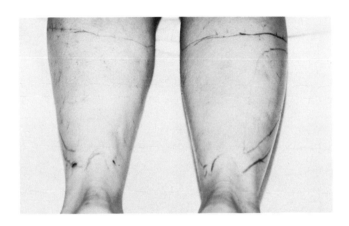

Figure 14–29. This intraoperative photograph shows the postoperative left leg with markings and incision visible.

The calves and ankles, which should be approached with the utmost care and circumspection, can be reached from incisions made on either side of the posterior ankle and occasionally below the knees (Figs. 14–29, 14–30). Although the knees are relatively easy to correct, calves and ankles require an exacting technique with 2-, 3-, and 4-mm blunt-tip cannulae and have no margin for error. Although an open-end cannula can be used, it is fraught with dangers if the surgeon is not thoroughly familiar with its use.[205] The postoperative edema in these areas can be particularly prolonged and can be helped if the patient wears support stockings for several weeks after the surgery. The patient should massage the postoperative edematous ankles, starting 7 to 10 days after the surgery, to prevent that edema from turning into fibrosis, which would require another procedure. Liposuction of the knees and calves can be performed with the patient either prone or supine. If the prone position is found more comfortable for the surgeon, the patient's feet should reach beyond the foot of the table. In this position, dorsiflexion of the foot can be helpful in allowing the cannula to be parallel with the skin (see Figs. 15–13, 15–14).

The anterior aspect of the legs can also be corrected with liposuction using the same exacting technique as for calves and ankles (Fig. 14–31). Severe lipodystrophy of legs can be helped, if not totally corrected, by careful liposuction of the area (Fig. 14–32).

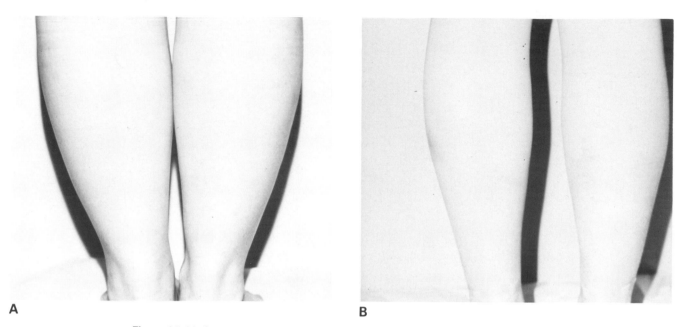

A   B

Figure 14–30. Compare these pre- and postoperative views of the same patient shown in Fig. 14–29.

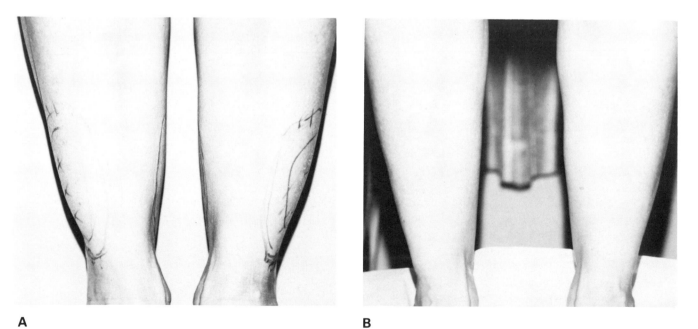

A   B

Figure 14–31. The results of liposuction on localized deposits of fat on anterolateral legs are shown. Liposuction below the knees should not be attempted by an inexperienced surgeon.

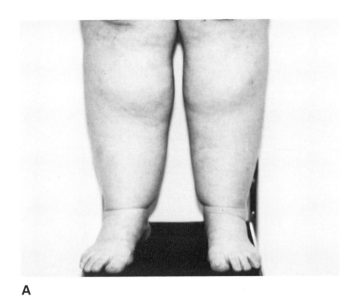

A

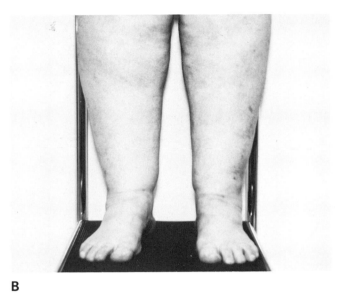

B

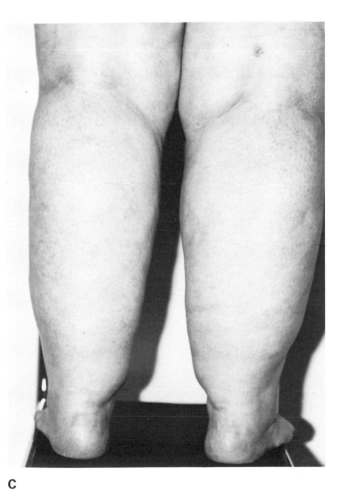

C

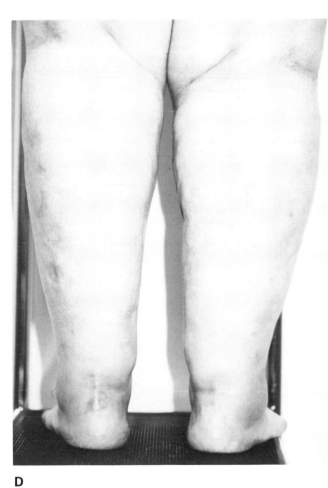

D

Figure 14–32. Severe lipodystrophy of the legs was helped considerably by liposuction surgery. As in the case demonstrated in Figure 14–10, this type of liposuction surgery is performed for functional as well as aesthetic reasons.

LIPOMAS

Lipomas, some of which may be huge (Figs. 14–33 to 14–35), are removed by the same technique.[96,97,128,206,207] The smaller lipomas can be quite fibrous; the liposuction may have to be complemented by surgical excision, although often one can use the cannula to tease out the fibrous tissue that may envelope or divide the fat (Fig. 14–36).

Lipomas in the anterior or posterior shoulder should be removed through a small axillary incision (Fig. 14–37). Lipomas greater than 4 to 5 cm in diameter can be easily removed with liposuction and are excellent lesions on which the practitioner can gain experience with liposuction.

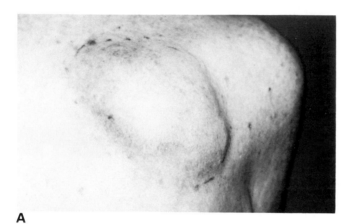

A

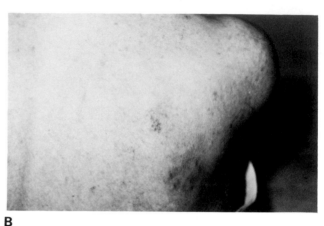

B

Figure 14–33. A lipoma on the back of a 74-year-old woman was removed with liposuction. This was the first experience I had with liposuction.

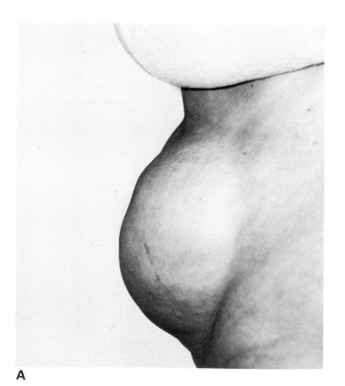

A

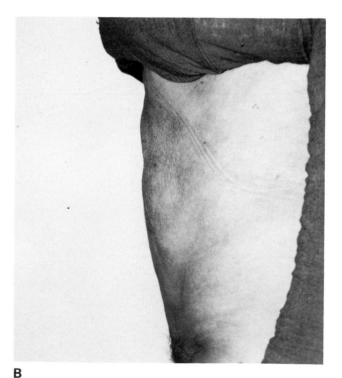

B

Figure 14–34. Pre- and postoperative views are shown of a large abdominal lipoma removed through an umbilical incision, in which 1800 ml of almost pure fat was obtained (see Fig. 5–11).

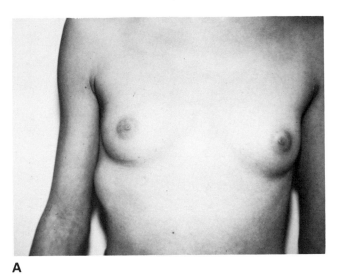

A

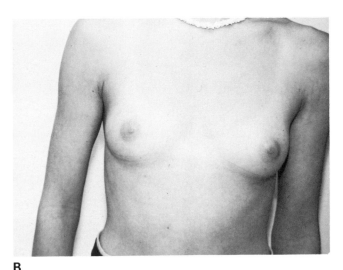

B

Figure 14–35. This case involved a lipoma in a 15-year-old girl who would not wear a bathing suit prior to liposuction.

Figure 14–36. The capsule surrounding or dividing the lipoma can be removed by teasing it out with the cannula.

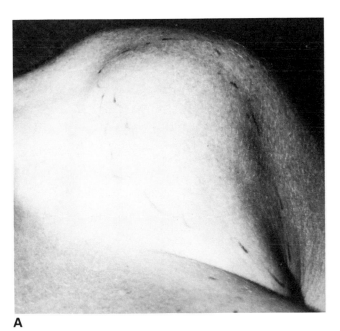

A

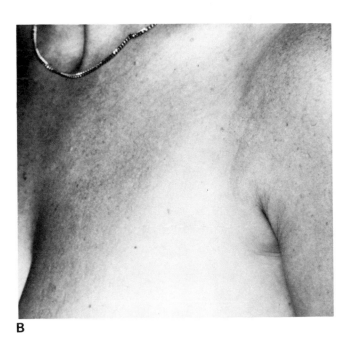

B

Figure 14–37. This lipoma of the anterior shoulder was removed through a ¼-inch incision made in the anterior axillary line.

111

Taping and Postoperative Care

15

The technique of taping the operated area is essential to a good final result. Use of the French Elastoplast (3 and 4 inch width) is highly recommended.* The tape redrapes the skin and its subcutaneous fat, particularly on the abdomen, to its proper position when the healing is complete. In addition, the tape acts as a compressive dressing, which prevents edema and shifting of fluids into the third space. The lack of complications during almost 5 years' experience with this technique in my own patients can be partly attributed to the careful application of the French tape, which helps diminish this loss of fluid within the body. A garment, no matter how tight fitting, cannot duplicate the compression given by properly applied tape.

TAPING

Face and Neck

The application of tape after liposuction of the melolabial fold should ensure pressure along the raised fold. This can be accomplished by applying a rolled 2 × 2-inch gauze or piece of compressible sponge[195] to the fold, over which Elastoplast is applied (Fig. 15–1). A piece of tape placed perpendicular to the fold, exerting traction toward the preauricular area, should be applied prior to the pressure dressing described above; it will help maintain the flattening effect of this dressing.

A device recently developed to maintain a constant pressure in this area[208] consists of two pads attached to elastic bands leading to a headpiece. This

device may be a valuable addition to facial liposuction surgery (Fig. 15–2).

Similarly, liposuction of the chin should be followed by the application of strips of Elastoplast 2 to 3 cm wide from the midline of the undersurface of the chin. Traction should be applied laterally to prevent vertical folds from forming in the midline. These folds can be prevented by leaving a 1-cm space in the midline.[194] This dressing is gradually built up (Fig. 15–3).

When liposuction of the jowls is performed, the strips of tape should exert traction from the undersurface of the chin, just beyond the mandibular ramus, upward to the temporal area. The cephalad end of the tape should be anchored at this point (Fig. 15–4).

Arms

Taping of the posterior arms should be done in a crisscross fashion. It should never be done in a circumferential manner (Fig. 15–5).

Chest

The chest can be taped from the sternal area laterally. However, because most men are rather hirsute in this area, an elastic binder may be preferable. An elastic vest is commonly employed (Fig. 15–6). An elastic binder may be applied over the vest.

Flanks

The tape should be applied over the entire area on which liposuction was performed. A wide binder will usually be adequate in a man (Fig. 15–7).

---

*Sold by Medicalex Co. of France and distributed in the United States by C.P. Medical Surgical, Newport Beach, Calif.

113

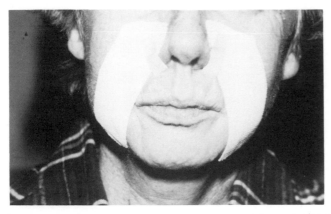

Figure 15–1. Tape is applied after liposuction of the melolabial fold through the external lateral nostril entry.

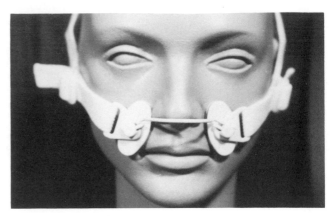

Figure 15–2. The pressure dressing device to be used on melolabial folds is shown.

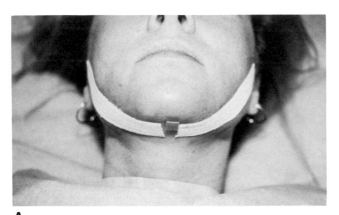

A

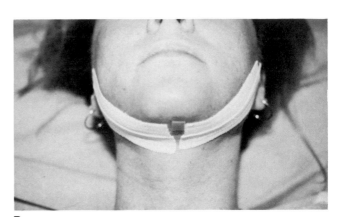

B

Figure 15–3. The correct application of tape after liposuction of the chin and jowls will ensure a good result.

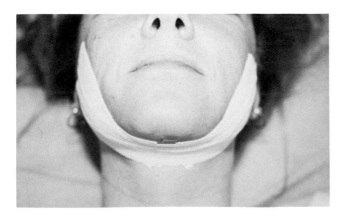

Figure 15–4. Taping follows liposuction of the chin and jowls.

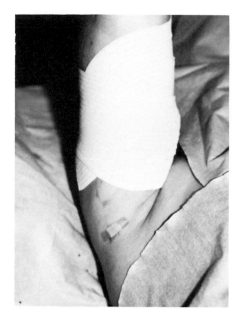
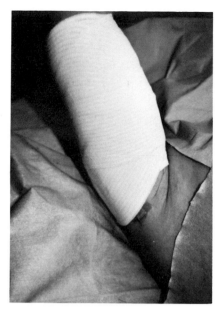

Figure 15–5. Taping of posterior arms, always in crisscross fashion, is gradually built up in thickness.

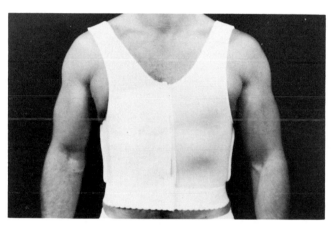

Figure 15–6. An elastic vest may be preferable in a hirsute man following liposuction for gynecomastia. *(Courtesy of Dunhill Medical, Inc.)*

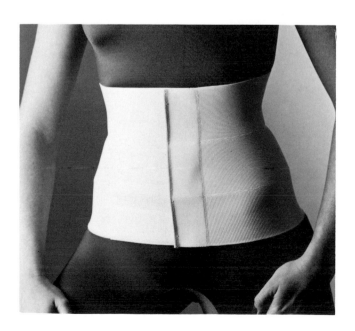

Figure 15–7. An elastic binder may be preferable in a man following liposuction of his flanks. *(Courtesy of Caromed International, Inc.)*

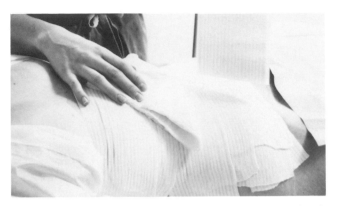

Figure 15–8. The epigastric area, susceptible to bleeding, is reinforced with a triangularly folded ABD pad, which fits between the ribs.

Abdomen

In the epigastric area, after some tape is applied, a Combine Dressing Pad folded in a triangular shape can be used to achieve greater compression and prevent edema and hematoma (Fig. 15–8). Taping of the abdomen requires observational experience. The French Elastoplast is applied upward in a crisscross fashion so that the skin and its subcutaneous fat is redraped in the proper position when healing is complete. Additional tape is applied to create a castlike dressing. There should be no wrinkles or defects in the tape (Fig. 15–9).

Application of an open 4 × 4-inch gauze over the pubic area can prevent adhesion of the tape to the

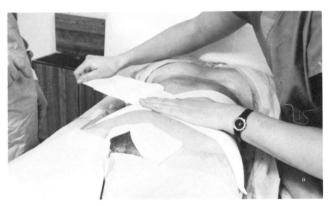

A

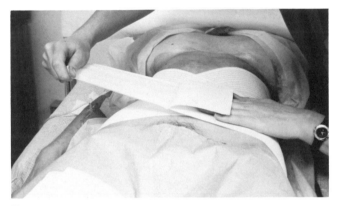

B

Figure 15–9. Taping of the abdomen and thighs requires observational experience. The French Elastoplast is applied from down upward such that the skin and its subcutaneous fat is redraped in the proper position when healing is complete. Additional tape is applied to create a castlike dressing. There should be no wrinkles or defects in the tape.

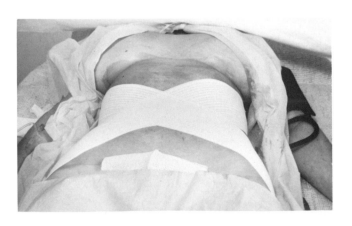

Figure 15–10. Adherence of tape to the pubic hair can be prevented by an open 4 × 4-inch gauze.

pubic hair, which is not appreciated by the patient, either during normal movement or when the tape is removed (Fig. 15–10).

Thighs and Hips

Compressive tape is applied on thighs by first using a long strip from the mid-lateral thigh over the trochanteric area, to the posterolateral aspects of the hip, and over to the lumbar area. This is followed by tape applied over the upper and posterior aspect of the thigh. The upper edge of this tape should be in

the gluteal fold; the lateral end is brought laterally and upward over the trochanter. This type of taping is continued until the entire area that underwent liposuction is covered by a castlike compressive dressing. This taping procedure also requires observational experience (Fig. 15–11).

Knees

The first tape to be applied to the knees should be in a crisscross fashion. This is carefully built up, ensuring good healing and preventing edema (Figs. 15–12, 15–13).

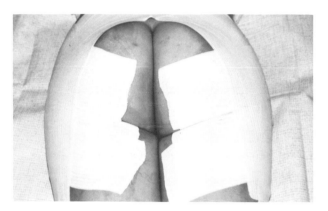

Figure 15–11. After liposuction of the thighs, the tape is applied from down upward to redrape the skin and the subcutaneous fat in the desired position.

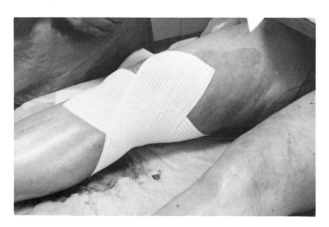

Figure 15–12. The first tape to be applied to the knees should be in a crisscross fashion.

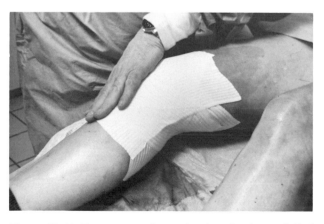

Figure 15–13. To prevent edema, the dressing is carefully built up after the initial crisscross tape is applied.

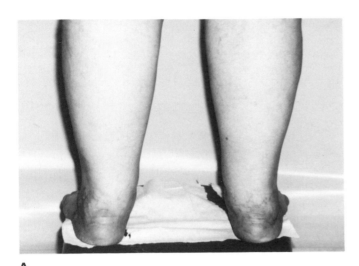

A

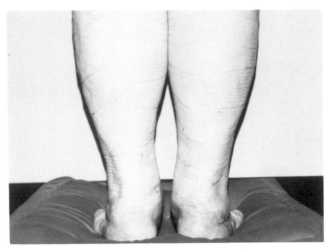

B

Figure 15–14. Before and after photographs of a patient who underwent liposuction of her calves and ankles are shown. The postoperative view shows the temporary markings left by the elastic bandage.

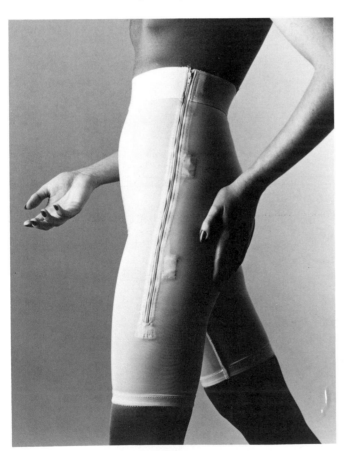

A

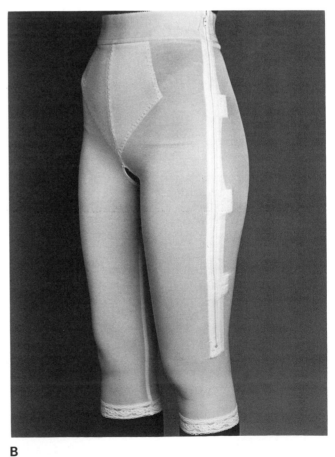

B

Figure 15–15. Support garments are worn by the patient for 4 to 6 weeks following liposuction surgery. The patient starts wearing the garment after the compressive tape is removed. *(Courtesy of Caromed International, Inc.)*

Calves and Ankles

Following liposuction of the calves and ankles, an elastic bandage should be applied in a crisscross fashion. Two ABD pads are then folded lengthwise (or two sanitary napkins) and placed on each side of the Achilles tendon just posterior to the malleoli and on each side of the gastrocnemius muscle. A second elastic bandage is then applied, also in a crisscross fashion (Fig. 15–14). These reinforced dressings will maintain the newly acquired contour and definition of the Achilles tendon by preventing edema, which can turn into permanent fibrosis.

REMOVAL OF TAPE

The tape is removed in 1 week; this is facilitated by the patient taking a hot bath (with bath oil) just before returning to the office. The patient should be warned of possible ecchymoses and edema. A support garment is worn for 4 to 6 weeks afterward (Fig. 15–15). Garments especially designed for post-liposuction wear are made by several manufacturers. In men who may have a hirsute body, an appropriate binder can be used instead of tape.

PHYSICAL EXERCISE AND MASSAGE

For details concerning what type of physical exercise and the time it should be commenced, see Chapter 8, which also discusses massage of the operated areas. If a physical therapist is available, more sophisticated massage may be employed.[188]

ULTRASOUND THERAPY

An ultrasound unit generates sound waves, which have a therapeutic effect on the tissue traumatized by liposuction surgery. Because it increases peripheral blood flow as well as the permeability of biologic membranes, ultrasound promotes increased absorption of intercellular fluids, with consequent reduction of edema. It also helps resolve the bruising and softens scar tissue, thus speeding up the patient's recovery.[189,190] The patient should undergo ultrasound therapy two to three times per week for approximately 6 weeks.

Autologous Fat Transplantation

16

The subject of autologous fat transplantation is discussed as related to the scope of this textbook, that is, fat transplantation by injection for the correction of facial and body contour.

HISTORY

Autologous fat transplants have been used to correct various skin defects for almost a century, since Neubauer transplanted fat from the upper arm into a depressed facial scar.[209] A number of workers have since published their results on using free fat grafts for aesthetic reasons in reconstructive, orthopedic, gynecologic, and other types of surgery. The early history of autologous fat transplantation was recounted by Neuhof, who also put forth his theory, albeit inaccurate, concerning the fate of the grafted adipocytes.[210] Further work on the histology, technique, and fate of the free fat grafts was done by Gurney, who also reported on the results achieved by his more immediate predecessors.[211]

It was Peer who brought the technique of free fat transplantation into the modern era.[212] In a series of most elegant in vivo studies (on experimental animals and human subjects), Peer demonstrated in a sequential manner the histology and fate of the transplanted fat.[22,23,213] Free fat grafts to the face,[214,215] the periorbital area,[216,217] and for reconstruction of the breasts[210,218] were reported in subsequent years.

With the advent of modern liposuction surgery and its dramatic success in recent years, research on fat metabolism, obesity, and the possible medical indications of liposuction received tremendous impetus. In 1984 Illouz discussed the possibility of using fat extracted with liposuction surgery by injecting it in certain skin defects.[219–221] The technique of fat extraction and transplantation by injection has advanced considerably, even though the first known injection of transplanted fat took place more than half a century ago.[222]

Macrolipoextraction (or liposuction) led to microlipoextraction and microlipoinjection.[223–225] In 1986 the first comprehensive manual on this subject was published.[83]

Peer's work was continued by Johnson, who determined the fate of the injected fat as well as the optimum quantity to be injected.[226] Bircoll and Johnson as well as Giorgio Fischer pioneered augmentation mammoplasties by transplantation of fat.[227,227a,227b]

HISTOLOGY

Contradictory reports stress the necessity of using either small[209,210] or large[212] fat grafts in order to ensure the survival of the fat. Similarly, a survival rate for transplanted fat of as little as 30 percent[211] or as much as 80 percent[227] has been reported during the past century.

Microscopically, the fat extracted by liposuction appears 90 percent viable, assuming it is not traumatized either by handling or by high suction pressure, which may cause vaporization.[225,228] The aspirated material, depending on the technique used, has been found to contain mostly fat and some blood, serum, and liquefied fat. Histologic examination of the fat shows mostly normal adipocytes with well-preserved cellular membranes[229] (Figs. 16–1, 16–2). Most histologic examinations performed on grafted fat have been on transplanted solid adipose tissue, demonstrating the transplanted fat undergoing inhibition of donor site plasma 72 hours after transplantation and inosculation of neovasculature into the graft 1 week post-transplant.[22,23,217,230]

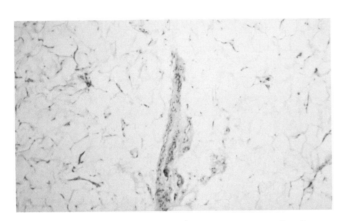

Figure 16–1. Histological examination of extracted fat demonstrates normal morphology of adipocytes (magnification × 20).

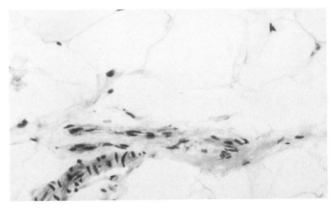

Figure 16–2. Normal cellular and nuclear morphology of extracted fat is shown (magnification × 100).

Only recently have we had the opportunity to examine fat grafted by microlipoinjection.[224,226,231]

In the case of transplanted "liquid" (injected) fat, the surviving cells may act as an implant with islands of fat cells floating in the interstitial tissue.[224,228,232] Illouz, in 1985, described his clinical impression of the injected transplanted fat as being lipoma-like.[224] A year later, adipose tissue freed by trauma has been found to survive in situ for 15 years as free viable lipomas.[233]

In the case of transplanted solid fat, the islands of surviving tissue develop into capsules formed by either leukocytes or connective tissue. Circulation in the fat capsules is evident a few days after transplantation and occurs by progressive anastamosis of the blood vessels of the host and graft tissue.[22,217]

Johnson found that the cellular reaction to the injected grafts and the formation of cysts within the grafts were related to the size of the transplant. Macrocysts and relatively little reaction were associated with small grafts, whereas macrocysts and a greater cellular host reaction occurred with larger grafts.[226]

CLINICAL INDICATIONS

Autologous fat transplantation can be used whenever recontouring of the face or body is indicated:[234,235]

I. Depressed scars
   A. Postsurgical
   B. Post-traumatic
II. Aging skin with loss of supportive tissue
   A. Glabellar furrows
   B. Upper lip
   C. Melolabial folds
   D. Hollow cheeks
   E. Dorsa of hands
III. Aesthetic enhancement
   A. Cheek augmentation
   B. Chin augmentation
   C. Breast augmentation
IV. Congenital defects
   A. Hemifacial atrophy
   B. Soft tissue defects of the body

Depressed Scars

Acne scars are difficult to correct with fat transplantation unless they are undermined to free the fibrous tissue retracting the skin. The ice pick scars cannot be corrected with this method.

1. Postsurgical scars may have to be treated first with pretunneling to free the scar tissue from the overlying skin (see Large Subdermal Defects, p. 126).
2. Post-traumatic scars caused by loss of tissue, particularly fat, are easily correctable with fat transplantation.

Aging Skin with Loss of Supportive Tissue

1. Glabellar furrows also may have to be undermined before being corrected. This can be accomplished, simply and effectively with a myringotomy knife.[226]
2. The upper lip responds well to autologous fat transplantation, assuming the skin is thin and loosely mobile. This indicates a good plane into which the fat can be injected. However, this procedure will not correct very fine wrinkles, particularly those adjacent to the vermilion. Ancillary methods may have to be used for the eradication of fine wrinkles.
3. Melolabial folds may respond if they show the same physical characteristics as those mentioned for the upper lip.
4. Hollow cheeks respond well to autologous fat transplantation. One must be particularly careful not to inject the fat below the subcutaneous fatty layer. Injury to blood vessels and nerves present in this area can bring about disastrous results.
5. Transplanted fat by microlipoinjection can be used to fill in and smooth out the thin and wrinkled skin present on the dorsa of hands, particularly between the tendon sheaths. Special attention should be given to the superficial veins, which can be easily broken. Here, in particular, injection upon withdrawal is important to prevent insertion of fatty material within the veins.

Aesthetic Enhancement

1. Cheek augmentation is very successful, assuming the fat is injected in the proper plane and location.
2. Chin augmentation works equally well; because of the pressure of the skin against the mentum (and its fat), this area may have to be injected more than once.
3. Breast augmentation should be undertaken with the utmost care. The morphology of the breast must be such (small, nonpendulous) that the injected fat will give the optimum result. Mammography and a history free of familial breast cancer is essential, not because the fat transplantation may cause a malignancy, but because the physical change occurring in the nonsurviving fat may cause radiographic and clinical findings indistinguishable from breast cancer.

Congenital Defects

1. Hemifacial atrophy has been treated with both solid and injected transplanted fat.[22,235,231]
2. Certain people are born with anomalies of their soft tissue contour, which may or may not be associated with bone deformities. In either case, autologous fat transplantation offers an alternative previously unavailable to these patients. Staged transplantation may be indicated when the defects are large (Figs. 16–3, 16–4). This may prevent either the formation of macrocysts within large transplants or their complete absorption and change into fibrous tissue.

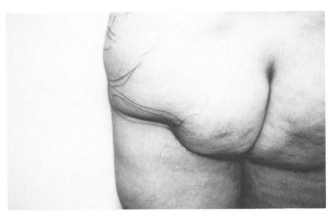

Figure 16–3. A congenital soft tissue defect prior to autologous fat transplantation is shown.

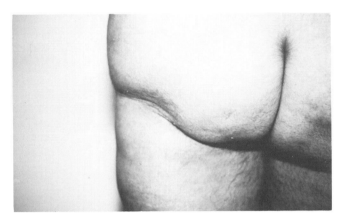

Figure 16–4. This postoperative view shows the appearance of the tissue after the first transplantation.

HARVESTING THE FAT

As with any surgical procedure, strict asepsis must be maintained, particularly because the harvested fat will be reinjected in a different location, leaving open the possibility of infection in both the donor and the recipient sites. Antibiotic therapy before the extraction of fat, by either the micro- or macrotechnique, may help prevent such infection because the tissue to be reinjected will have been perfused with an antibiotic.

Microlipoextraction

Microlipoextraction refers to aspiration of less than 20 ml of fat for the purpose of reinjecting it. The donor site is anesthesized either with lidocaine (Xylocaine) solution or with neocryoanesthesia, using saline solution chilled to 2C or an ice cube applied directly to the skin.[225] The site of entry in the skin may also have to be anesthesized with a few drops of Xylocaine. If a microcannula is used rather than a

needle, a 2-mm incision is made with a size 11 blade. Often the only anesthesia necessary for microlipoextraction is for the skin of the donor site alone.[234]

If the quantity of fat necessary for microlipoinjection is assumed to be less than 20 ml, a syringe can be attached to the distal end of a suction tube.[225] By moving the syringe with the needle (a 14-gauge has been found to be the most satisfactory) back and forth within the fatty tissue (with the suction on), sufficient fat can be obtained to fill one or more syringes (Fig. 16–5). Lowered suction pressure will avoid excessive trauma to the extracted fat and thus obtain more viable cells.[236] There is no particular advantage of using a syringe connected to an aspirating unit, however, because the flow of fat into the syringe is impeded by the small diameter of the syringe tip, whether Luer-Lok or not. Similarly, more fat cannot be aspirated through a large-diameter cannula attached to a syringe, also because of the small opening of the syringe tip.

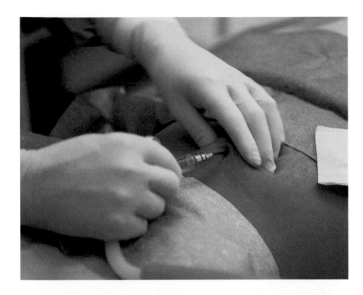

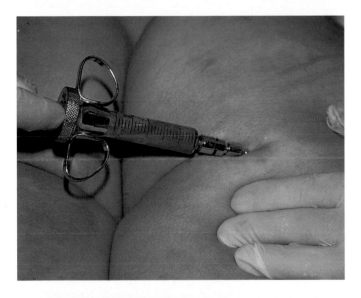

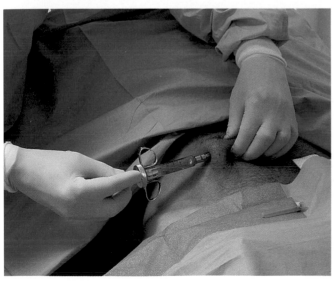

Figure 16–5. After removal of the plunger, the syringe can be connected to an aspirator unit by insertion of appropriate-size tubing into its proximal end.

Figure 16–6. Pulling the plunger will create sufficient vacuum within the syringe to aspirate a moderate quantity of fat. The gluteal fold, thigh, and hip are ideal donor sites.

Figure 16–7. Although the abdomen can be quite tender after microlipoextraction, it can serve as a donor site.

A further simplification of this method of extracting fat is the use of the syringe without a suction unit. By inserting the needle into the adipose tissue and pulling the plunger of the syringe, sufficient vacuum can be created to aspirate fat (Figs. 16–6 to 16–8).[236] An 18-gauge needle threaded through the plunger will maintain the negative pressure within the syringe (Fig. 16–9). The use of a sharp needle to extract the fat may traumatize the tissue, and in particular the capillary network, which may cause more bleeding than desired. Although a needle with a larger bore will obtain parcellar fat, similar to a liver biopsy, it will also induce more bleeding because of the larger cutting surface of its tip.

The use of a microcannula (approximately 2 to 3 mm in diameter with three openings on one side near the tip) may minimize the trauma and permit the aspiration of fat with less blood (Figs. 16–10, 16–11). Variations on this type of microcannula are constantly being devised in the hope of not only minimizing trauma but increasing the yield of extracted fat as well (Fig. 16–12).

If a needle is used for extraction, one has only to reinject the fat in the chosen site. If a microcannula is used, it is simply removed from the collecting syringe and replaced with an appropriate needle, which is then used for injection.

Whether one uses a sharp needle or a blunt-tip cannula (or variations on its design), the trauma to the fat induced by the pistonlike movement can cause the aspiration of an excessive amount of unwanted fluid, depending on where the fat is taken from, the type of cannula used, and the technique of the operator. This fluid consists of serum, fat derived from broken adipocytes, blood, and possibly local anesthetic if it was previously injected (Figs. 16–13, 16–14). The more fluid in the syringe, the more overcorrection is necessary, with temporary unsightly results. In addition, if 60 percent of the extracted tissue consists of fat and 80 percent of that fat is viable, only 40 percent of the extracted tissue can be injected as an implant. Furthermore, only about 60 percent of that tissue may survive, which means

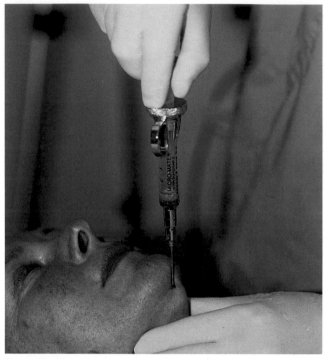

Figure 16–8. With this technique of mini liposuction, small deposits of fat can be removed from the undersurface of the chin.

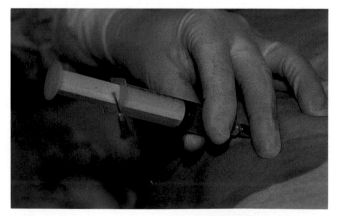

Figure 16–9. After sufficient traction is applied on the plunger to create a vacuum within the syringe, an 18-gauge needle is inserted through the plunger. This will maintain its position and thus the vacuum within the syringe. This method is applicable only if a plastic syringe with this particular type of plunger is used.

that in effect a syringe containing a fair amount of serum and liquefied fat may deliver only approximately 30 percent of its volume as fat that will "take." This is too small a percentage and requires too much overcorrection to make this system, without modification, a practical one.

My method of reducing the material to be injected to practically pure fat is to place the fat-filled syringe into a centrifuge. The syringe is then spun for a few seconds at the desired revolutions per minute, and the serum, blood, and liquefied fat will col-lect in the dependent part of the syringe, that is, near the tip. Taking the needle off will cause this extraneous fluid to drain rapidly into a piece of gauze or a container, and the fat can be injected immediately (Fig. 16–15).

For fat transplantation to cheeks (10 to 15 ml), melolabial folds (10 to 15 ml), upper lip (5 to 10 ml), glabella (5 ml), or hands (10 to 20 ml), the extraction of fat with a syringe using either a needle or a micro-cannula is both feasible and practical using the above method.

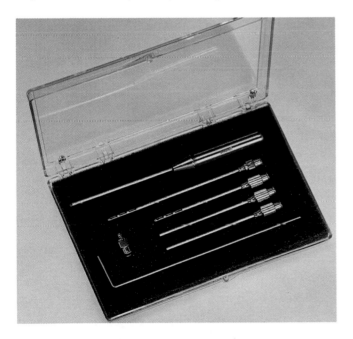

Figure 16–10. The Asken Microextractor Kit consisting of a tunneler and four microextractors (all of which can also be used for reinjection of fat). *(Courtesy of Robbins Instruments, Chatham, N.J.)*

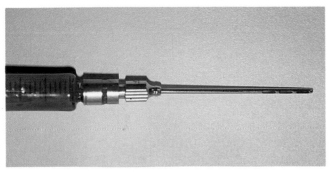

Figure 16–11. The microextractor is attached to the syringe.

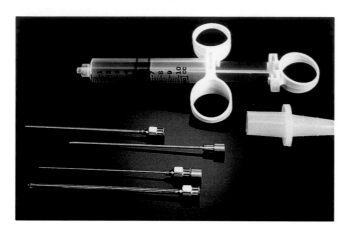

Figure 16–12. A control-syringe with different microextractors is shown. *(Courtesy of Byron Medical, Tuscon, Ariz.)*

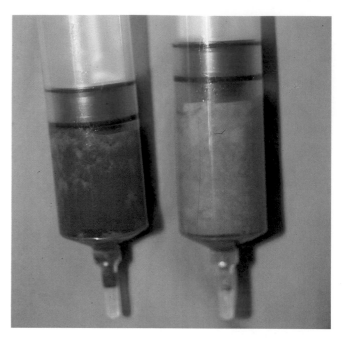

Figure 16–13. Extracted fat: fat with excessive amount of fluid in it (*left*) and good quality fat (*right*).

Figure 16–14. Syringe after having been spun in the centrifuge showing the middle layer of fat representing approximately 60 percent of extracted material.

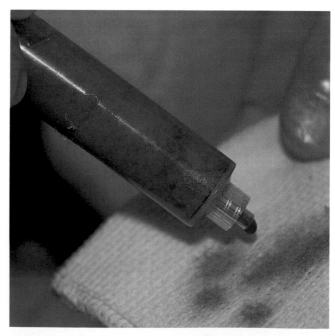

Figure 16–15. By choosing the proper donor site and microcannula, a minimal amount of fluid will be present in the extracting syringe. After centrifuging the fat, the excess fluid can be removed by draining it into a gauze.

Macrolipoextraction

Macrolipoextraction refers to the extraction of larger quantities of fat than used for correction of a few deep wrinkles (i.e., more than 20 ml). Considering the volume of fluid present in the syringe that will not contribute to the correction of defects, a large quantity of fat must be extracted using other methods.

A collecting device consisting of a plastic container within which a sterile filter trap is present has been found to be the most practical method for this technique.* The floor of the filter trap consists of a rubber plunger disk which, when pushed with the provided rod, empties the trapped fat directly into a syringe. This avoids possible contamination of the extracted fat. This in-line collecting device is attached at one end to a suction unit and at the other end to the cannula of choice, preferably one 3 to 4 mm in diameter (Fig. 16–16).

The tip of this cannula should be as atraumatic as possible (bullet shaped). After the area is anesthetized (either skin or fat or both), the cannula is inserted into the adipose layer and the suction unit is started and brought to approximately two thirds of the maximum negative pressure.

The surgeon then performs the same piston-like movement as in routine liposuction. The aspirated fat will be caught within the filter trap and the serum, local anesthetic, and liquefied fat will pass into the collecting jar. When the trap is nearly full, the proximal end of the plastic casing attached

*In Line Collection Set, Berkeley Medevices, Inc., Berkeley, Calif.

to the suction unit is detached from the collecting tube, and the unit is switched off. Only then is the cannula removed from the adipose layer. The collecting device is reattached to the collecting tube, and the suction unit is started again. By allowing a flow of air to pass through the trap containing the suctioned fat, excess fluid present in it will drain into the collection bottle. One may want to apply intermittent occlusion at the distal end of the collecting device, which will cause further drainage of excess fluid. The fat is then aseptically deposited into the injecting syringe (Figs. 16–17 to 16–20). If this full collecting device is placed into a container for further drainage, very little liquid will be observed in it.

By contrast, if the collecting device is placed within a container without first allowing the air to go through it, as described above, considerable drainage will result (Figs. 16–21, 16–22).

There are many other methods of harvesting fat for transplantation, but all seem to have certain disadvantages that limit their practicality. This is especially true when many physicians perform autologous fat transplantation in their offices on an outpatient basis.

The main disadvantage of injecting the aspirated fat directly from the collecting syringe is the large amount of serum and liquefied fat, which reduce the actual implanting material to considerably less than what is extracted.

The fat can also be collected in a sterile container, strained, placed in a syringe, and in-

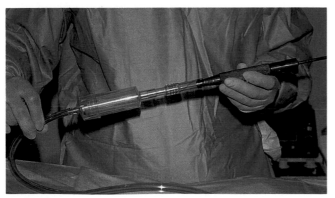

Figure 16–16. The In Line Collecting Set is shown attached to the cannula and the suction tubing.

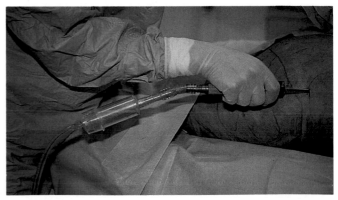

Figure 16–17. The filter basket is partially filled with fat within the collection container.

jected.[237] This system seems awkward because it requires ancillary personnel, more steps, and more handling of the fat than desirable. It also carries with it the serious possibility of bacterial contamination if not performed in a sterile environment, such as an operating room.

A technique used to estimate the amount of fat removed by liposuction employs a closed stockinette attached to the inlet tube of the collecting jar.[238] This collecting device could also be used for the fat intended for transplantation. Again, the chief objection to this technique is the extra handling of the fatty tissue, hence the increased possibility that foreign material from the stockinette (or any gauze strainer) will enter the fat and be injected along with it. This may cause a foreign-body reaction and imperil the entire transplanted fat graft.

Another method of collecting the fat involves the use of a sterile jar with two metal tubes through its stopper. One tube is attached to the cannula used for extracting the fat and deposits the suctioned material into the bottle. The other tube is attached to the hose leading to the suction unit. This second tube is long, with its opening only a few millimeters from the bottom of the jar. It suctions off any liquid (blood, serum, anesthetic) accumulated at the bottom of the collecting bottle. The remaining fat in the bottle is then placed in syringes for reinjection.[231] Here, too, the manipulation of fat may injure the cells and imperil their viability.

The main advantages of my technique of micro- and macrolipoextraction are the minimal handling of the fat and the assurance that its sterility is maintained. In microlipoextraction, the aspirated fat stays in the collecting syringe, and its sterility is preserved. The minimal centrifuging necessary to separate the blood and serum from the fat does not injure the adipocytes.

In macrolipoextraction, the aspirated fat is similarly kept under sterile conditions either in the trap or in the syringe, into which it is atraumatically deposited. In addition, the simplicity of these two methods make them available to all physicians interested in this procedure.

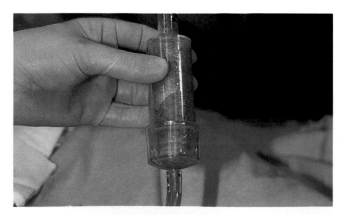

Figure 16–18. The filled filter basket shows the excess serum draining into the tube by simply letting the suction on for a few seconds.

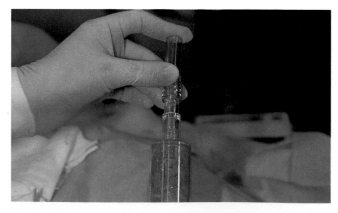

Figure 16–19. By intermittently occluding the inlet opening, the excess fluid will rapidly pass into the collecting bottle.

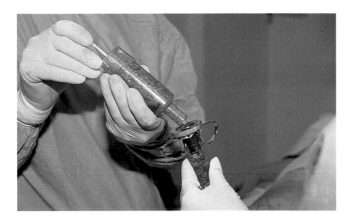

Figure 16–20. By using the plunger rod provided with this collection set, the fat is extruded directly into the syringe.

Figure 16–21. The passage of air through the collecting device will drain the excess fluid into the suction tube and very little fluid will be obtained by gravity drainage.

Figure 16–22. If the serum is not drained by letting the air through the collecting chamber and the filter basket, the trapped fat contains considerably more fluid which can be drained by gravity alone.

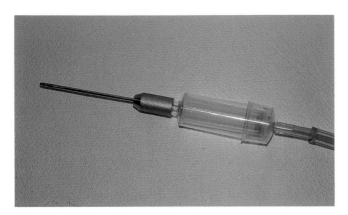

Figure 16–23. The Asken Harvester is attached to the Berkeley In Line Collection Set.

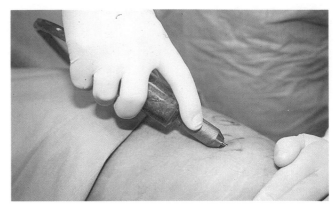

Figure 16–24. This intraoperative photograph shows lipoextraction using the Asken Harvester.

In micro- as well as in macrolipoextraction, a short 3-mm outside-diameter (OD) cannula with three oval apertures has been found to be an elegant as well as practical instrument.* This cannula is connected either to the in-line collection set or directly to the aspirating hose (Figs. 16–23 to 16–25). Its small size and practically atraumatic tip make it an ideal instrument for harvesting fat to be used for autologous transplantation.[239,240] The short length of this instrument will facilitate the extraction of fat because the flow through a tube is inversely proportional to its length: the shorter the instrument, the faster the flow of fat due to the greater pressure gradient within the tube (Poiseuille's law). Similarly, if the diameter of the aspirating tube is greater than that of a 14-gauge needle, the volume of suctioned fat is increased.

More recent work performed by Dolsky demonstrated a significant difference in the morphology and possible survival rate of the extracted adipocytes, as shown by electron microscopic examination and acid phosphatase activity, depending on the instrumentation used for harvesting and injecting the fat.[241] The damage incurred by the adipocytes is inversely related to the diameter of the instrument used to extract and inject them.

Similar conclusions were drawn by Campbell

Figure 16–25. Note the appearance of the extracted fat after the excess fluid has been drained using the method described in the text.

and Newman, who studied the integrity of the adipocytes by their ability to maintain their functions as measured by their glucose metabolism and fatty acid synthesis.[242]

---

*Asken Adipose Tissue Harvester, Robbins Instruments, Inc., Chatham, N.J.

Conclusion

It must be remembered that harvesting the fat for transplantation by using microlipoextraction even without an aspirating unit constitutes liposuction. By modifying the cannula and the pressure of the suction unit, macrolipoextraction for fat transplantation can be performed with the same minimal damage to the fat cells. In either case, the donor site has the potential for resulting irregularities, hematoma, seroma, and injury to the underlying tissue.

Fat can be extracted from any part of the body that has a sufficiently thick adipose layer. However, upper abdomen and flanks, which are more fibrous, are not ideal donor sites. Similarly, the lower abdomen seems to have more post-operative tenderness than do thighs and buttocks and should be avoided as a donor site if possible. The pubic area seems to be a particularly good donor site for fat intended for transplantation.[243]

Theoretically, fat should be transplanted from an area of less vascularization into one with greater vascularization. The fat should bear the trauma of transplantation somewhat better and may have a greater chance of "taking," once implanted into a site with more circulation than the area from which it was taken.

However, important consideration should be given to the preponderance of receptor type present on the donor adipocytes (see Fig. 2–7). Fat for transplantation should be taken from areas known to have the highest antilipolytic activity, such as the thighs (see Fig. 2–8). Thus, the transplanted fat, once it is vascularized and accepted by the organism as normal local tissue, has a greater chance of maintaining its volume, wherever it is implanted.[226]

The donor site should have the same type of compressive dressing to prevent hematoma and ex-

cessive edema as at any other site in which liposuction is performed. Similarly, for a defect corrected with transplanted fat, cold compresses should be applied, also to prevent edema, ecchymosis, and hematoma.

TREATMENT OF EXTRACTED FAT

Aspirated fat has been treated in various ways. In an effort to separate it from serum, blood, fibrous tissue, and injected anesthetic and to make it more viable after transplantation, it has been filtered, sifted, rinsed, washed, spun, and agitated. Some physicians believe that the aspirated fat should be rinsed with saline, whereas others maintain that lactated Ringer's is a more physiologic solution with which to wash the fat.

Aspirated fat has been treated with saline and heparin in the belief that they will increase its viability (see Fig. 16–28). Early work seemed to demonstrate the retardation or prevention of lysis of cultured fat cells, however, by perfusing them with insulin.[244,245]

The rationale of perfusing the aspirated cells intended for transplantation with insulin is further strengthened by the fact that the fat cells have specific insulin (anti-lipolytic) receptors[59,246] (see Fig. 16–30). One of the main reasons for loss of implanted (solid) fat is lysis[22,23]; any treatment to prevent or retard this lysis will increase the survival of transplanted fat. More recent work further confirmed the beneficial antilipolytic effect of insulin on these cells.[247–249]

In addition to the antilipolytic effect of insulin, vitamin E has been shown to participate in the metabolism of fatty tissue and has been used empirically to enhance the survival rate of the transplanted fat.[250]

If insulin is added to a perfusing solution, it is either in the ratio of 100 units of regular insulin to 1000 ml of solution (saline or Ringer's) or in a 1:1 ratio of one unit of insulin per 1 ml fat.[217] Furthermore, leaving the insulin in contact with the fat cells for approximately 1 hour at room temperature will presumably ensure its binding to the adipocytes cell membrane, possibly resulting in longer survival of the adipocytes. Campbell demonstrated that addition of insulin to extracted fat increases its glucose metabolism. This effect was increased by first washing the fat.[242]

No experimental work has yet been done to prove (or disprove) the antilipolytic effect of insulin on the injected transplanted fat or the resulting increased survival rate.

AUTOLOGOUS FAT TRANSPLANTATION

Transplantation of fat by injection implies that the consistency of the fat is such that it can pass through a needle. An injection of any substance, however, particularly a viscous one, requires anesthesia to the area of injection.

Nerve block anesthesia of the recipient site is preferable; if not possible, a minimal amount of anesthetic fluid should be injected in the recipient site to avoid distortion of the area to be corrected.

Large Subdermal Defects

An area that has lost its soft tissue due to trauma does not usually have scar tissue that can interfere with the injection of fat. When it does occur, the injection of fat can be performed with a 16-gauge needle using a 10-ml control syringe. A larger syringe will impede the flow of fat through the needle and will make the injection unnecessarily difficult. The length of the needle will be dictated by the area to be corrected. A large defect will require a longer needle, whereas a smaller defect can be corrected with a short needle. The length of the needle is important in minimizing the number of punctures in the skin. In addition to possible scarring, these openings may lead to extrusion of the fat injected under a certain amount of pressure. Consequently, injection of fat should be performed with the same principle as that of giving local anesthesia: once the needle is under the skin the injection is performed on withdrawing the needle and in a radial fashion. Injection upon withdrawal minimizes the possibility of intravascular injection of fat with the possible consequence of fat emboli, although it does not eliminate it (see Fig. 16–31).

If the defect has a great deal of scar tissue, it may have to be broken down with a fine tunneler (see Fig. 16–10) to allow for an even dispersal of the injected fat. This network of fine tunnels will permit the injected fat to flow through them, just as molten metal will flow into a sculpture mold when the lost wax method of casting is used. If there are several openings in the skin when these tunnels are filled with fat, the excess will extrude through these openings just as the molten metal flows out when the mold is filled. Unless the tunneling is performed at several layers, simply filling them with fat will not correct a depressed defect. Consequently, to prevent the loss of transplanted fat, a minimum of skin perforations should be made.

Furrows, Deep Wrinkles, or Small Defects

Tunneling is rarely necessary, although previous undermining with a myringotomy knife may be helpful.[226] Injection upon withdrawal is always practiced for reasons mentioned above (Figs. 16–26 to 16–29).

Injection of glabellar furrows should be performed away from the nose, and toward the forehead, to avoid a possible sudden spurt of fat into the periorbital area (Fig. 16–30). Occasionally one may have to bend the needle slightly to accommodate the position of the hand on the patient's face or nose.

Melolabial folds respond well to autologous

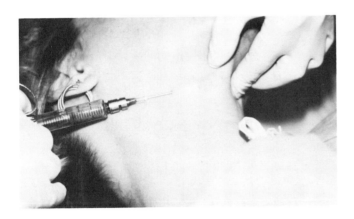

Figure 16–26. Autologous fat injection of scar in the neck is demonstrated. Note the position of the needle parallel with the skin.

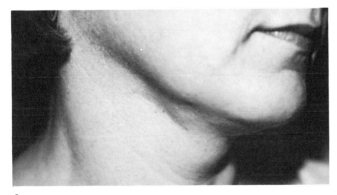

A

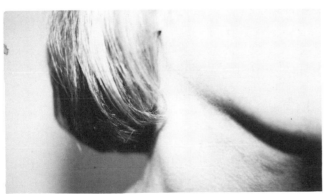

B

Figure 16–27. Pre- and postoperative views of the patient seen in Figure 16–26.

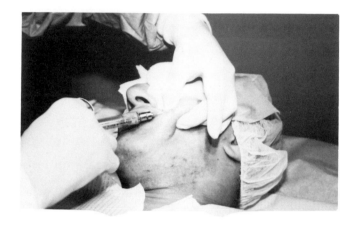

Figure 16–28. Fat transplantation of facial scars is shown.

A

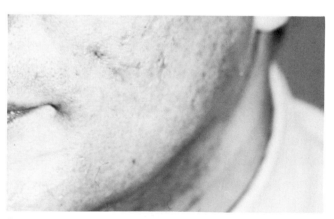

B

Figure 16–29. Deeply retracted scars may require several fat transplantation procedures because of the pressure of the surrounding fibrous tissue on the transplanted fat.

fat transplantation, provided there is loss of tissue and the skin is not thick and indurated. The point of entry is usually lateral to the commissures of the mouth and within a crease. The needle is threaded exactly under the furrow or wrinkle, and the fat is injected upon withdrawal. Injection of fat lateral to furrows will only accentuate them by increasing the mass of surrounding tissue (Fig. 16–31).

The upper lip can also be filled with transplanted fat, assuming the skin is loose and mobile on the underlying tissue. The fat is injected from the melolabial fold in a radial fashion, with only one puncture of the skin. Fine wrinkles, however, will not respond to fat transplantation (Fig. 16–32).

Both the upper lip and the melolabial folds should be anesthesized by a regional nerve block prior to the injection of fat. The upper gingival sulcus is first treated with dental Xylocaine ointment, after which it is infiltrated in several areas with 1 percent Xylocaine with epinephrine, using approxi-

mately 5 ml of anesthetic for the entire sulcus (Fig. 16–33). The anterior-superior alveolar nerve, which derives from the infraorbital nerve (a branch of the maxillary division of the trigeminal), is thus anesthesized without distorting the lip.

After a 10- to 15-minute wait, the fat transplantation can proceed. Similarly, the anesthesia of the infraorbital nerve will permit cheek augmentation with fat transplantation.

Augmentation of the cheeks or of the chin can be done either through the gingival sulcus or through the skin.[2,34] If liposuction of the chin has been performed previously, the same incision can be used to inject the fat into the mental area (Figs. 16–34, 16–35). If augmentation of the cheeks over the zygoma is desired, the intraoral route is simple and avoids puncturing the skin (Figs. 16–36, 16–37). For hollow cheeks, the external route is preferable, as the needle can be kept parallel to the skin (Figs. 16–38 to 16–40).

Injection of fat in the face should be done with particular care, to avoid its dispersal in the peri-

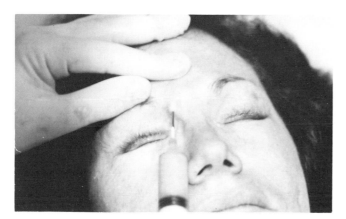

Figure 16–30. Injection of glabellar furrows should be performed towards the forehead to prevent possible injection of fat into the periorbital area.

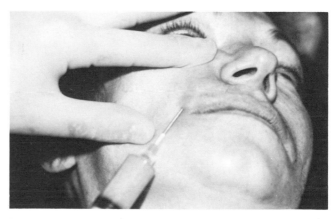

Figure 16–31. Fat transplantation into melolabial folds should be performed exactly under the furrow and not lateral to it. Note the index finger on the inferior orbital ridge.

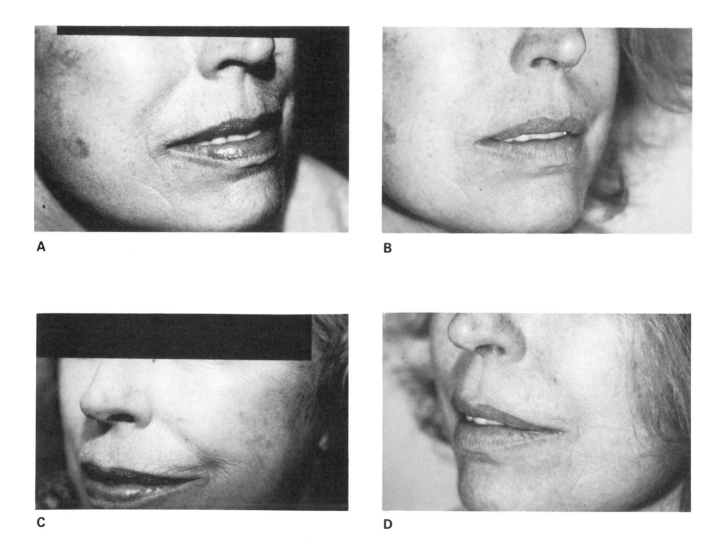

Figure 16–32. Pre- and postoperative views of patient who underwent autologous fat transplantation to upper lip and melolabial folds.

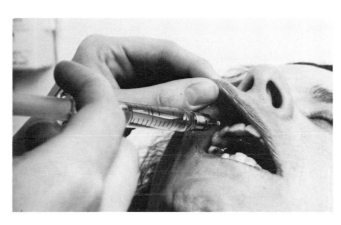

Figure 16–33. Regional block anesthesia is preferable to local infiltration prior to autologous fat transplantation, because distortion of the recipient area is avoided.

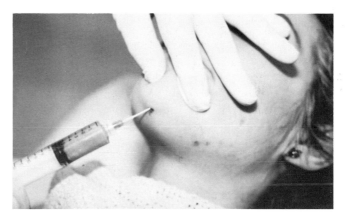

Figure 16–34. Transplantation of fat for chin augmentation through an incision made for liposuction of the undersurface of the chin.

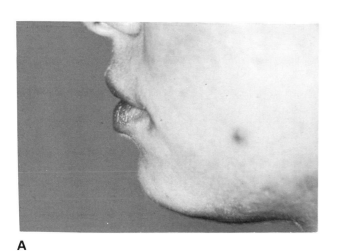

A

B

Figure 16–35. Pre- and postoperative photographs of the patient shown in Figure 16–34.

orbital area. One finger should apply pressure on the infraorbital ridge whenever the fat is injected upward into the cheeks. This possible complication can be minimized by injecting from the temporal area inferiorly and away from the eye. The area in which the fat has been injected should be molded to eliminate skin irregularities (Fig. 16–41).

The dorsa of hands can be similarly injected, filling in and smoothing out the depressed areas between the tendon sheaths. Here in particular one must avoid puncturing the veins, which can produce long-lasting ecchymoses. These can counteract any beneficial aesthetic effect derived from the injection. Injection on withdrawal will add to the success of this procedure.

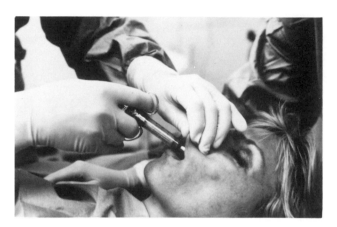

Figure 16–36. Intraoral route of fat transplantation for augmentation of cheeks.

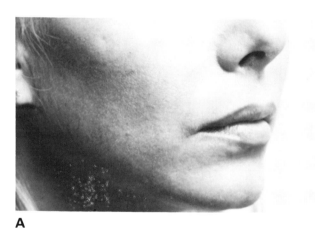

A

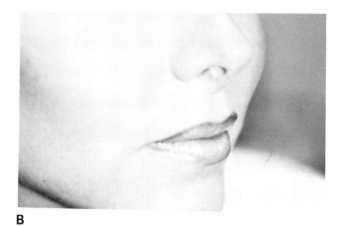

B

Figure 16–37. Pre- and postoperative photographs of patient seen in Figure 16–36.

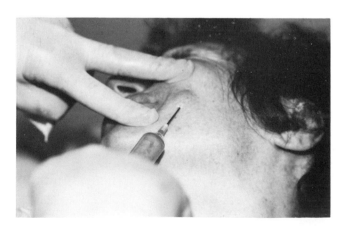

Figure 16–38. Position of the needle when correcting a defect on the face is shown.

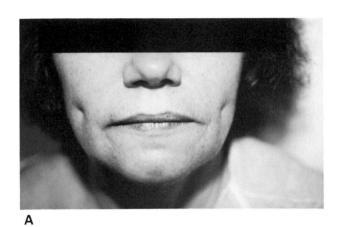

A

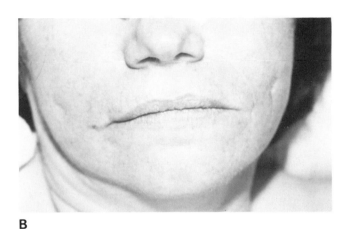

B

Figure 16–39. Pre- and postoperative views of a patient who desired correction of her exaggerated dimples (see Fig. 16–38).

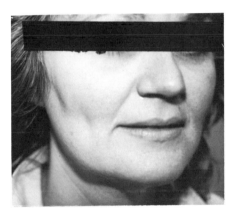

A

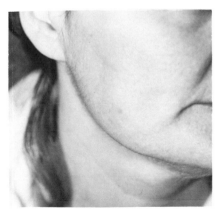

B

Figure 16–40. Partial correction of surface irregularities of the cheek is demonstrated.

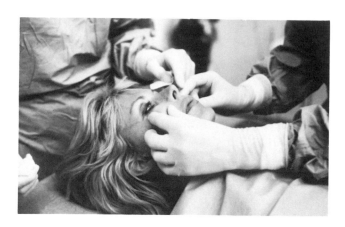

Figure 16–41. The area where the fat has been injected should be molded to eliminate possible surface irregularities.

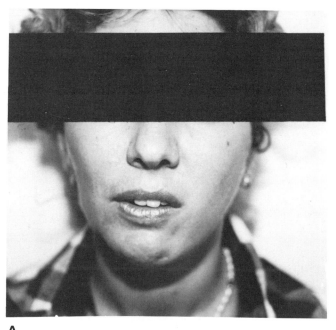

A

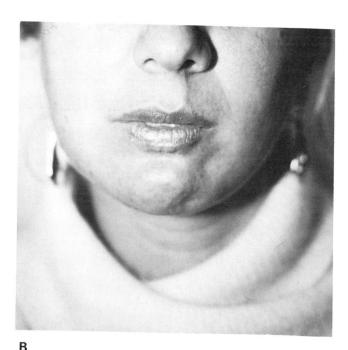

B

Figure 16–42. Pre- and postoperative photographs of patient with hemifacial atrophy demonstrate correction using autologous fat transplantation.

Congenital defects such as hemifacial atrophy can be corrected with autologous fat transplantation (Fig. 16–42). The sequence of injections should be performed first proximally and then distally, that is, away from the operator, to avoid extrusion of fat through the openings in the skin from previous injections.

Breast augmentation by fat transplantation has been reported for at least 60 years.[210,218] Only during the past few years, however, has fat obtained by liposuction suction been used for breast augmentation by injection.[226,227] The breast must have the proper morphology, to obtain the best aesthetic results. A detailed history, a careful physical examination, and a mammogram should be performed prior to surgery.

Local anesthesia of breast tissue can be difficult; the patient will require adequate sedation and analgesia prior to the injection of the anesthetic. A 20-gauge spinal needle is first inserted under the breast and the anesthetic is injected slowly by advancing the needle, and not on withdrawing it. Aspiration prior to injection is essential to avoid intravascular injection of the anesthetic. The addition of hyaluronidase will facilitate diffusion of the anesthesia. Waiting a few seconds between the initial insertion and the forward motion of the needle will permit the area to become somewhat anesthesized, and there will be less discomfort. The anesthetic should first be injected between the breast tissue and the muscle. After a few minutes, some anesthesia can be injected into the breast tissue, this time by withdrawing the needle. An 18-gauge needle will facilitate injection of the anesthetic within the glandular tissue.

Injection of autologous fat should be performed after a 10- to 15-minute wait. It should start in the lower levels of the breast (close to the muscle) and gradually move to the more superficial areas in a radial fashion. The injection should be performed upon withdrawal, to avoid depositing a large amount of fat in one location, which will only cause its lysis or the formation of macrocysts. A long needle can be used to inject the fat, 16-gauge or larger. The use of an atraumatic microcannula may produce less bleeding, although resistance to it upon insertion will be greater (Fig. 16–43). Tunneling prior to fat injection may facilitate its dispersion within the breast tissue (Fig. 16–44). Similarly, the fat can be injected in the retromammary plane, between the breast parenchyma and the muscle. Injecting the fat in different levels to preserve its filamentous character and, consequently, adequate circulation to all the fat injected, may be difficult in such a case, however.

Because of the resistance to the inserting needle or blunt-tip cannula, the tip of the plastic syringe can bend or break in response to the motions made by the operator's hand during insertion of the needle. Consequently a 10-ml glass control syringe is preferable to a plastic one (Fig. 16–45).

Breast augmentation by autologous fat trans-

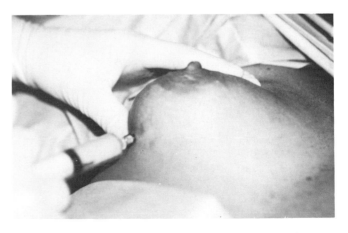

Figure 16–43. Injection of autologous fat into the breast can be done through the inframammary fold and in a radial fashion using the Asken microcannula.

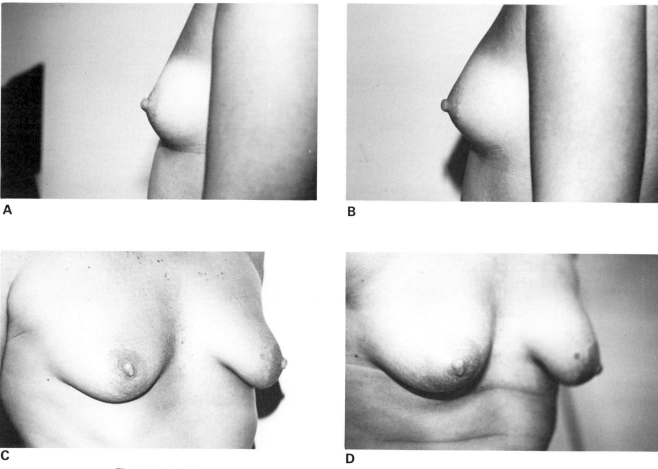

Figure 16–44. Moderate breast augmentation can be achieved with fat transplantation.

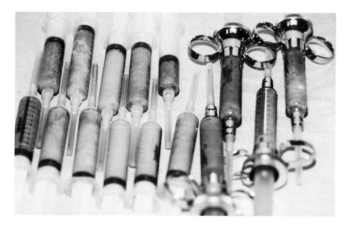

Figure 16–45. For ease of injection, the aspirated fat should be placed into 10-ml syringes before the onset of transplantation. Glass syringes are preferable for injection of fat in the breast.

plantation should be performed with the utmost care and with the patient's understanding that it is a relatively new procedure. Although there are no known cases of breast cancer induced by fat transplantation, a nidus of cancerous cells may be present prior to the fat injection. The lasting effect of breast augmentation by this method is unknown. The patient should be made thoroughly aware of these facts, and a signed informed consent and release should be obtained prior to attempting this procedure.

Before signing the consent, the patient should understand the nature of autologous fat transplantation. As it is a relatively new procedure, there are conflicting reports concerning survival of transplanted fat. There are so many variables in the technique used for extracting it as well as for injecting it, however, that no meaningful interpretation can be made of the data available today.

Following is the consent form used by the author for this procedure.

Consent for Autologous Fat Transplantation

Patient: _____

Date: _____   Time: _____

I hereby authorize _____ and/or his associates to perform autologous fat transplantation to my _____. I understand the fat will be taken from my _____. I fully understand that this procedure has limited application. No guarantee or assurance has been given to me as to the results that may be obtained and how long the transplanted fat will last. I am aware that the practice of medicine and surgery is not an exact science, and I acknowledge that no guarantees have been made to me as to the results of the operation or process.

Dr. _____ has discussed in detail with me the information that is briefly summarized below:

I. *Nature and Purpose of Autologous Fat Transplantation*

An autologous fat transplant involves the removal of fat from a donor site, usually the hips, thighs or abdomen, which is then injected into the desired recipient site to correct certain defects or signs of aging.

In autologous fat transplantation, local anesthesia with sedation may be used. The donor site is taped with a compressive tape and ice-cold compresses are applied to the recipient site which may be swollen for a few days after the surgery.

II. *Risks*

I understand that among the known risks are bruising, numbness, lumpiness, and swelling.

I am aware that, in addition to the risks specifically described above, there are other risks such as loss of blood and infection that may accompany any surgical procedure as well as injury to the nerves which may lead to temporary numbness.

I recognize that, during the course of the operation, unforeseen conditions may necessitate additional different procedures than those set forth above. I therefore further authorize and request that the above named surgeon, his assistants, or his designees perform such procedures as are in his professional judgment necessary and desirable.

III. *Anesthesia*

I understand that local anesthesia will be used with sedation. I consent to the administration of local anesthesia with sedation by or under the administration of _____.

I am aware that risks are involved with the administration of anesthesia, both local anesthesia and sedation, such as allergic or toxic reactions to the anesthetic, and respiratory or cardiac arrest.

IV. *Alternatives to Autologous Fat Transplantation*

Alternative methods of correcting skin defects do exist such as collagen or silicone injections. These methods were fully explained to me.

V. *Informed Consent*

I have had sufficient opportunity to discuss my condition and proposed surgery with _____, and all my questions have been answered to my satisfaction. I understand that fat transplantation is a relatively new procedure, and no guarantees or promises have been made to me regarding how long the transplanted fat will last. I believe that I have adequate knowledge on which to base an informed consent to the proposed treatment.

VI. *Photographs*

I consent to be photographed before, during, and after the treatment and that these photographs shall be the property of the above doctor and may be published in scientific journals or for scientific reasons.

VII. *Cooperation*

I agree to keep Dr. _____ and the staff informed of any change in my permanent address, and I agree to cooperate with them in my aftercare.

_____   _____
Patient or Legal Guardian   Witness

Overcorrection by at least 30 percent is necessary to conpensate for the loss of fat.

Postoperative care of the recipient site is simple: cold compresses for a few hours, oral antibiotics, and oral or parenteral steroids to reduce the postoperative swelling. Although statistically there is no proof of its efficiency, clinically the steroids seem to help reduce postoperative edema.

The donor area should be treated with the same care as any area upon which liposuction has been performed.

The use of autologous fat transplantation for the correction of certain skin defects and signs of aging is an alternative to other means available today for treating these conditions. However, fat transplantation is not a substitute either for collagen or for silicone injection. It is just another method in the armamentarium of the cosmetic surgeon for rejuvenating the face and body and for the correction—even though it may be temporary—of certain defects.

References

1. Caver C: The unpublished history of liposuction. Presented at the Fourth Annual Scientific Meeting of the American Society of Lipo-Suction Surgery, Los Angeles, January 1985.
2. Schrudde J: Lipexeresis in the correction of local adiposity. In Proceedings of the First Congress of the International Society of Aesthetic Surgery, Rio de Janeiro, 1972.
3. Schrudde J: Lipexeresis in the correction of local adiposity Int J Aesthetic Plast Surg, 1972.
4. Schrudde J: Lipexeresis as a new type of aesthetic plastic surgery. Transactions of the Second Congress of the International Confederation of Plastic and Reconstructive Surgery, Madrid, May 1973.
5. Schrudde J: Lipexeresis as a new type of aesthetic plastic surgery Aesthetic Plast Surg 4:215, 1980.
6. Fischer A, Fischer GM: Revised technique for cellulitis fat: Reduction in riding breeches deformity. Bull Int Acad Cosmetic Surg 2:40, 1977.
7. Kesselring UK, Meyer R: A suction curette for removal of excessive local deposits of subcutaneous fat. Plast Reconstr Surg 62:305, 1978.
8. Illouz YG: Un nouveau traitement chirurgical sur les lipodistrophies localisées Presentations at the Société Française de Chirurgie Esthétique, June 1978 and 1979.
9. Illouz YG: Une nouvelle technique pour les lipodistrophies localisées. Rev Chir Esth Langue Fr 6(19):3, 1980.
10. Fournier PF: Facelift: New SMAS and defattening procedures. Presentation sponsored by the Sacramento Society of Otolaryngology and Maxillo-Facial Surgery, Lake Tahoe, February 1982.
11. Fredericks S (chairman): Report of the Commission on Surgical Suction Lipectomy by the Ad-Hoc Committee on New Procedures. Executive Committee Meeting of the American Society of Plastic and Reconstructive Surgery, San Francisco, January 1983.
12. Grazier F: Discussion: Suction-assisted lipectomy, suction lipectomy, lipolysis and lipexeresis. Plast Reconstr Surg 72(5):620, 1983.
13. Hetter GP, et al: Nomenclature, in Hetter GP (ed): Lipoplasty: The Theory and Practice of Blunt Suction Lipectomy. Boston, Little, Brown, 1984, pp. 65–76.
14. Newman J: Lipo-suction surgery: Past, present, future. Am J Cosm Surg 1:19, 1984.
15. Virchow R: On malignant tumors of adipose tissue. Virchows Arch [A] 11:281, 1857.
16. Wasserman F: Development of adipose tissue, in Reynold AE, Cahill GF (eds): Adipose Tissue. Baltimore, Williams & Wilkins, 1965, p 87.
17. Simon G: Genèse et structure du tissu adipeux chez l'homme. Acta Anat 48:232, 1962.
18. Wells HG: Adipose tissue, a neglected subject. JAMA 114:2177, 1940.
19. Smahel J, Clodius L: The blood vessel system of free human skin grafts. Plast Reconstr Surg 47:61, 1971.
20. Pearl RM, Johnson D: The vascular supply to the skin: An anatomical and physiological reappraisal. Part I. Ann Plast Surg 11:99, 1983.
21. Knittle KJL, Timmers K, Ginsberg Fellner F, et al: The growth of adipose tissue in children and adolescents. Cross sectional and longitudinal studies of adipose cell number and size. J Clin Invest 63:239, 1979.
22. Peer LA: Loss of weight and volume in human fat grafts. Plast Reconstr Surg 5:217, 1950.
23. Peer LA: The neglected free fat graft. Plast Reconstr Surg 18:233, 1956.
24. Peer LA: The neglected "free fat graft": Its behavior and clinical use. Am J Surg 92:40, 1956.
25. Peer LA: Transplantation of Tissues. Vol II. Baltimore, Williams & Wilkins, 1959.
26. Hansberger FX: On the ability of transplanted fetal adipose tissue from rats to grow and develop. Virchows Arch [A] 302:640, 1938.
27. Vague VJ, Boyer J, Jubelin J, et al: Adipo-muscular ratio in human subjects, in Vague J, Denton R (eds): Physiopathology of Adipose Tissue. Amsterdam, Excerpta Medica, 1969.
28. Krotkiewski KM, Sjöström L, Bjönjtorp P, et al: Impact of obesity on metabolism in men and women. Importance of regional adipose tissue distribution. J

Clin Invest 72:1150, 1983.

29. Faust IM, Miller WH Jr: Effect of diet and environment on adipocyte development. Int J Obes 5:593, 1981.

30. Hirsch J, Batchelor B: Adipose tissue cellularity in human obesity. Clin Endocrinol Metab 5:299, 1976.

31. Krotkiewski M, et al: Adipose tissue cellularity in relation to prognosis for weight reduction. Int J Obes 1:395, 1977.

32. Oliver MF: The vulnerable myocardium. Lancet 2:560, 1973.

33. Kammel WB, et al: Obesity, lipids and glucose intolerance: The Framingham study. Am J Clin Nutr 32:1238, 1939.

34. Kannel WB, Gordon T: Obesity and some physiological and medical concommitants, in Bray GA (ed): Obesity in America. Washington, DC, National Institutes of Health 1979.

35. Kannel WB, Castelli WP: Prognostic implications of blood lipid measurements, in Freis JF, Ehrlich GE (eds.): Prognosis, Bowie, Md., Charles Press Publications, 1980.

36. Hirsch J, Gallian E: Methods for determination of adipose cell size in man and animals. J Lipid Res 9:110, 1968.

37. Sims EA, et al: Endocrine and metabolic effects on experimental obesity in man. Recent Prog Horm Res 29:457, 1973.

38. Fain JN, Garcia-Sauiz JA: Adrenergic regulation of adipocyte metabolism. J Lipid Res 2:945, 1983.

39. Bjorntrop P, Ostman J: Human adipose tissue. Dynamics and regulation. Adv Metab Disord 5:277, 1971.

40. Green H, Kelinde O: Spontaneous heritable changes leading to increased adipose conversion in 3T3 cells. Cell 7:105, 1976.

41. Bjorntrop P, et al: Quantification of different cells in the epidermal fat pad of the rat. J Lipid Res 20:97, 1979.

42. Roncari DAK, et al: Exaggerated replication in culture of adipocyte precursors from massively obese persons. Metabolism 30:425, 1981.

43. Bjorntrop P, et al: Differentiation and function of rat adipocyte precursor cells in primary culture. J Lipid Res 21:714, 1980.

44. Hirsch J: Hypothalamic control of appetite. Hosp Pract 19(2):131, 1984.

45. Amatruda JM, et al: Hypothalamic and pituitary dysfunction in obese males. Int J Obes 6:183, 1982.

46. Von Gierke E: About the metabolism of adipose tissue. Verh Dtsch Ges Pathol 10:182, 1906 [In German].

47. Wood RW: Potpourri of lipid tissue: Literature peregrinations. Aesthetic Plast Surg 8:247, 1984.

48. Illouz YG: Surgical remodeling of the silhouette by aspiration lipolysis or selective lipectomy. Aesthetic Plast Surg 9:7, 1985.

49. Smith U: Adrenergic control of human adipose tissue lipolysis. Eur J Clin Invest 10:343, 1980.

50. Smith U, Digirolamo M, Blohm G, et al: Possible systemic metabolic effect of regional adiposity in a patient with Werner syndrome. Int J Obes 4:153, 1980.

51. Arner P, Ostman G: Relationship between the tissue level cyclic AMP and the fat cell size of human adipose tissue. J Lipid Res, 19:613, 1978.

52. Arner P, Ostman J: Importance of the cyclic AMP concentrations on the rate of lipolysis in human adipose tissue. Clin Sci Mol Med 59:199, 1980.

53. Engfeldt P, Arner P, Wahrenberg H, et al: An assay for beta-adrenergic receptors in isolated human fat cells. J Lipid Res 23:715, 1982.

54. Arner P: Human adipose tissue: Function, development and metabolism, in Hetter GP (ed): Lipoplasty: The Theory and Practice of Blunt Suction Lipectomy. Boston, Little, Brown, 1984, pp 41–48.

55. Engfeldt P, Arner P, Ostman J: Studies on the regulation of phosphodiesterase in human adipose tissue in vitro. J Clin Endocrinol Metab 56:501, 1983.

56. Vaughn M: The production and release of glycerol by adipose tissue incubated in vitro. J Biol Chem 237:3354, 1962.

57. Le Boeuf B, et al: Effect of epinephrine on glucose up-

take and glycerol release by adipose tissue in vitro. Proc Soc Exp Biol Med 102:527, 1959.

58. Czech M: Insulin action. Am J Med 70:142, 1981.

59. Pedersen O, et al: Insulin receptor binding and receptor mediated insulin degradation in human adipocytes. Diabetologia 20:636, 1981.

60. Larner J: Insulin-mediator—Fact or fancy? J Cyclic Nucleotide Res 8:289, 1982.

61. Mendelson CR, et al: Growth factors suppress and phorbol esters potentiate the action of dibutyryl adenosine 3', 5'-monophosphate to stimulate aromatase activity of human adipose stromal cells. Endocrinology 118:3, 1986.

62. Thorner MO: Hypothalamic releasing hormones: Clinical possibilities. Hosp Pract 21(12):63, 1986.

63. Temourian B, Kroll SS: Subcutaneous endoscopy in suction lipectomy. Plast Reconstr Surg 74:5, 1984.

64. Kesselring UK: Regional fat aspiration for body contouring. Plast Reconstr Surg 11:610, 1983.

65. Kesselring UK: Body contouring with suction lipectomy. Clin Plast Surg 11:3, 1984.

66. Chajcir A: Suction curettage lipectomy. Aesthetic Plast Surg 7:195, 1983.

67. Hetter G: Optimum vacuum pressures for lipolysis. Aesthetic Plast Surg 8:1, 1983.

68. Temourian B, et al: Suction lipectomy: A review of 200 patients over a six year period and a study of the technique in cadavers. Ann Plast Surg 11(2):93, 1983.

69. Temourian B, Fischer JB: Suction curettage to remove excess fat for body contouring. Plast Reconstr Surg 68:50, 1981.

70. Hetter GP: Physics and Equipment, in Hetter GP (ed): Lipoplasty: The Theory and Practice of Blunt Suction Lipectomy. Boston, Little, Brown, 1984, pp 119–135.

71. Illouz YG: Aspiration: Résultats à long terme et commentaires. Rev Chir Esth Langue Fr 10(41):7, 1985.

72. Illouz YG, Fournier PF: Collapsochirurgie et remodelage de la silhouette, Cah Chir 12(46):1, 1983.

73. Fischer G: Personal technique of liposuction. Presented at the Second International Congress of Aesthetic Surgery, Paris, May 1984.

74. Fischer G: Historia de la liposucción. Presentation at the International Course of Liposuction and Aesthetic Surgery, Sociedad Andaluza de Cirurgia Plastica, Reparadora y Estetica, Malaga, Costa del Sol, June 1984.

75. Fischer G: The Technique of Liposuction. Presented at the Lipolysis Symposium, Lipolysis Society of North America, Las Vegas, October 1984.

76. Fischer G: Body sculpturing after nine years of liposuction. Presented at the International Congress of the International Society of Aesthetic Surgery, Tokyo, May 1985.

77. Fournier PF, Otteni FM: Traitement des lipodystrophies localisées par aspiration. La technique sèche, in J Faivre (ed): Chirurgie Esthétique 1981. Paris, Maloine, 1982, pp 59–78.

78. Fournier, PF: Recent advances in lipoplasty. Presented at the Second International Congress of the International Society of Aesthetic Surgery, Tokyo, May 1985.

79. Fournier PF: Réflections sur la lipoplastie. Rev Chir Esth Langue Fr 10(41):23, 1985.

80. Fournier PF, Otteni FM: Honeycomb lipectomy or liposuction-dissection in body sculpturing—The dry procedure. Annual Meeting of the American Society of Plastic Reconstructive Surgery, Honolulu, Hawaii, October 1982.

81. Fournier PF, Otteni FM: Plasties abdominales hier et aujourd'hui—Dermolipectiomés et collapsochirurgie. Cah Chir 20:(whole issue), 1984.

82. Otteni FM, Fournier PF: La chirurgie de la silhouette: de la chirurgie classique à la collapsochirurgie. Rev Chir Esth Fr 8(32):13, 1983.

83. Asken S: A Manual of Liposuction Surgery and Autologous Fat Transplantation Under Local Anesthesia, ed 2. Irvine, Calif., Keith C. Terry and Associates, 1986, pp 23–26.

84. Illouz YG: Reflexious après 4 ans et demi d'experience et 800 cas de ma technique de lipolyse. J Chir Esth Montreuil 6(24):12, 1981.

85. Illouz YG: Une nouvelle technique pour les lipodystrophies localisées: La lipoectomie sélective ou lipolyse. In J Faivre (Ed.) Chirurgie Esthétique 1981–1982. Paris, Maloine, 1982, pp 79–92.

86. Illouz YG: Body contouring by lipolysis: A 5 year experience with over 3000 cases. Plast Reconstr Surg 72:59, 1983.

87. Fournier PF, Otteni FM: Lipodissection in body sculpturing: The dry procedure. Plast Reconstr Surg 72:598 1983.

88. Fournier PF: Cryoanesthesia and cryolipoplasty. Presentation at the Lipolysis Society of North America Inc., Annual Meeting, Las Vegas, October 1984.

89. Fournier PF: Cryoanesthésie et cryolipolastie—néocryoanesthésie et neocryolipoplastie. Rev Chir Esth Langue Fr 10(41):27, 1985.

90. Hetter GP: The effect of low dose epinephrine on the hematocrit drop following lipolysis. Presented at the First Annual Meeting of the Lipolysis Society of North America, Las Vegas, Nevada, 1983.

91. Asken S: Comparison of dry and wet liposuction techniques. Presented at the Fourth Annual Scientific Meeting of the American Society of Cosmetic Surgeons, Los Angeles, January 1985.

92. Hetter GP: Surgical technique, in Hetter GP (ed): Lipoplasty: The Theory and Practice of Blunt Suction Lipectomy. Boston, Little, Brown, 1984, pp 137–154.

93. Martin N: Introduction to Lipolysis to the United States: A three-year experience, in Hetter GP (ed): Lipoplasty: The Theory and Practice of Blunt Suction Lipectomy. Boston, Little, Brown, 1984, pp 37–39.

94. Dolsky RL: Fluid and blood dynamics of lipo-suction surgery. Presented at the Third Annual Scientific Meeting of the American Society of Lipo-Suction Surgery, Los Angeles, January 1984.

95. Dolsky RL: Computerized analysis of instrument, techniques and solutions. Presented at the Second World Congress on Liposuction Surgery, The Graduate Hospital, Philadelphia, June 1986.

96. Asken S: Liposuccion de la silhouette et du visage en ambulatoire en utilisant l'anésthesie locale. Rev Chir Esth Fr 10(40):5, 1985.

97. Asken S: Liposuction surgery as performed under local anesthesia in an outpatient surgical facility. J Int Soc Aesthetic Surg 3:1, 1985.

98. Bestler J: Liposuction on obese patients—A preliminary report. Presented at the American Society of Liposuction Surgery. the Fourth Annual Scientific Meeting of the American Society of Cosmetic Surgeons, Los Angeles, January 1985.

99. Krulig E: Liposuction and obesity. Presented at the Second World Congress on Lipo-suction Surgery, The Graduate Hospital, Philadelphia, June 1986.

100. Burke K: Results of lipo-suction on fat distribution and eating behavior. Presented at the Second World Congress on Lipo-suction Surgery, The Graduate Hospital, Philadelphia, June 1986.

101. Gendler J: Physiologic justification for lipo-suction surgery. Presented at the Second World Congress on Liposuction Surgery, The Graduate Hospital, Philadelphia, June 1986.

102. Chajchir A: Liposuction and obesity. Presented at the Second World Congress on Lipo-suction Surgery, The Graduate Hospital, Philadelphia, June 1986.

103. Tcheupdijian L: Metabolic aspects of human obesity. Presented at the Second World Congress on Lipo-suction Surgery, The Graduate Hospital, Philadelphia, June 1986.

104. Illouz YG: Illouz's technique of body contouring by lipolysis. Clin Plast Surg 11:3, 1984.

105. Stallings JO: The defatting of flaps by lipolysis, in Hetter GP (ed): Lipoplasty: The Theory and Practice of Blunt Suction Lipectomy. Boston, Little, Brown, 1984, pp 309–321.

106. Manders EK, et al: Elimination of lymphangioma circumscriptum by suction-assisted lipectomy. Ann Plast Surg 16:532, 1986.

107. Otteni FM: Intérêt de la technique française de lipoplastie avec aspiration dans la maladie de Launois-Bensaude. Rev Chir Esth Langue Fr 11(44):63, 1986.

108. Luscher NJ, et al: Lipomatosis of the neck (Made-

lung's neck). Ann Plast Surg 16:502, 1986.

109. Sonensheim H, Lepoudre C: Suction assisted lipectomy—A functional use in the neck. Am J Cosmetic Surg 2:42, 1985.

110. Hallock GG: Adjuvant use of suction lipectomy cannula for blunt dissection. Aesthetic Plast Surg 9:2, 1985.

111. Field LM: Adjunctive liposurgical debulking and flap dissection in neck reconstruction. J Dermatol Surg Oncol 12:9, 1986.

112. Winslow RB: The use of lipolysis for unusual diseases (treatment of congenital lymphedema of the lower extremity), in Hetter GP (ed): Lipoplasty: The Theory and Practice of Blunt Suction Lipectomy. Boston, Little, Brown, 1984, pp 323–329.

113. Temourian B: Face and neck suction-assisted lipectomy associated with rhytidectomy. Plast Reconstr Surg 7:628, 1983.

114. Newman J, Dolsky RL: Liposuction surgery: History and development. J Dermatol Surg Oncol 10:6, 1984.

115. Chrisman B, Field L: Facelift surgery update: Suction-assisted rhytidectomy and other improvements. J Dermatol Surg Oncol 10:7 1984.

116. Otteni FM: Le triple lifting cervicofacial. Rev Chir Esth Langue Fr 10(41):41, 1985.

117. Newman J, Nguyen A: Cervical Facial Liposuction in Ambulatory Surgery and Office Procedures, in Head and Neck Surgery. Orlando, Fl., Grune & Stratton, 1986, pp 149–183.

118. Fournier PF, Otteni FM: Plasties abdominales hier et aujourd'hui—Dermolipectomies et collapso chirurgie. Cah Chir 50:(whole issue), 1984.

119. Otteni FM, Fournier PF: Les ombilicoplasties dans la chirurgie esthétique de l'abdomen. Rev Chir Esth Langue Fr 10(39):31, 1985.

120. Dardour JC, Vilain R: Alternatives to classic abdominoplasty. Ann Plast Surg 17:3, 1986.

121. Baroudi R: Lipolysis combined with conventional surgery, in Hetter GP (ed): Lipoplasty: The Theory and Practice of Blunt Suction Lipectomy. Boston, Little, Brown, 1984, pp 277–293.

122. Vilain R, Dardour JC: Aesthetic surgery of the medial thighs. Ann Plast Surg 17:3, 1986.

123. Baroudi R: Body sculpturing. Clin Plast Surg 11:3, 1984.

124. Aiache AE: Lipolysis of the female breast, in Hetter GP (ed): Lipoplasty: The Theory and Practice of Blunt Suction Lipectomy. Boston, Little, Brown, 1984, pp 227–231.

125. Epstein LI: Buccal lipectomy. Ann Plast Surg 5:2, 1980.

126. Newman J, Nguyen A, Anderson R: Lipo-suction of the buccal fat pad. Am J Cosmetic Surg 3:1, 1986.

127. Asken S: Lipo-Suction under local anesthesia. Presented at the First World Congress and Scientific Meeting on Liposuction Surgery, Cannes, France, June 1984.

128. Field LM, Asken S, Caver CV, et al: Lipo-suction surgery: A review. J Dermatol Surg Oncol 10:7, 1984.

129. Asken S: Recent progress in aesthetic surgery. Presentation at the Third International Workshop of the International Academy of Cosmetic Surgery, Rome, Italy, May 1987.

130. Smahel J: Adipose tissue in plastic surgery. Ann Plast Surg 16:450, 1986.

131. Kesselring UK: Suction curette for removal of subcutaneous fat (letter to the editor). Plast Reconstr Surg 63:560, 1979.

132. Elam MV: Cobra cannula, in Liposuction: The Franco American Experience. Beverly Hills, Medical Aesthetics, 1965, pp 146–147.

133. Winslow RB: A pistol-grip cannula for lipo-extraction. Ann Plast Surg 14:1, 1985.

134. Tucker GT, Mather LE: Clinical pharmacokinetics of local anesthetics. Clin Pharmacokinet 4:241, 1979.

135. Meridy HW: Criteria for selection of ambulatory surgical patients and guidelines for anesthetic management—A retrospective study of 1553 cases. Anesth Analg 61:921, 1982.

136. Giles HG, et al: Influence of age and previous use on diazepam dosage required for endoscopy. Can Med Assoc J 118:513, 1978.

137. Mandelli M, et al: Clinical pharmacokinetics of diazepam. Clin Pharmacokinet 3:72, 1978.
138. Kortilla K, Linnoila M: Psychomotor skills related to driving after intramuscular administration of diazepam and meperidine. Anesthesiology 42:685, 1975.
139. Kortilla K, Linnoila M: Recovery and skills related to driving after intravenous sedation: Dose–response relationship with diazepam. Br J Anesth 47:457, 1975.
140. MacDonald RL, Barker JL: Enhancement of GABA mediated postsynaptic inhibition in cultured mammalian spinal cord neurons: A common mode of anticonvulsant action. Brain Res 167:323, 1979.
141. Tomichek RC, et al: Cardiovascular effects of diazepam-fentanyl anesthesia in patients with coronary artery disease. Anesth Analg, 61:217 1982.
142. Bailey, PL, et al: Anesthetic induction with fentanyl. Anesth Analg 64:48, 1985.
143. Allonen, H, et al: Midazolam kinetics. Clin Pharmacol Ther 30:653, 1981.
144. Smith MT, et al: The pharmacokinetics of midazolam in man. Eur J Clin Pharmacol 19:271, 1981.
145. Fragen RJ, et al: A water soluble benzodiazepine, Ro 21-3981, for induction of anesthesia. Anesthesiology 49:41, 1978.
146. Reves JG, et al: Comparison of two benzodiazepines for anesthetic induction: Midazolam and diazepam. Can Anesth Soc J 25:211, 1978.
147. Fragen RJ, Caldwell NJ: Recovery from midazolam used for short operations. Anesthesiology 53:S11, 1980.
148. Fragen RJ, Caldwell NJ: Awakening characteristics following anesthesia induction with midazolam for short surgical procedures. Arzneimittelforschung 31:2261, 1981.
149. McClure JH, et al: Comparison of the I.V. administration of midazolam and diazepam as sedation during spinal anesthesia. Br J Anesth 55:1089, 1983.
150. Beckett AH, Casey AF: Synthetic analgesics, stereochemical considerations. J Pharm Pharmacol 6:986, 1954.
151. Pasternak GW, Snyder SH: Identification of novel high affinity opiate receptor binding in rat brain. Nature 253:563, 1975.
152. Pert A, Yaksh T: Sites of morphine-induced analgesia in the primate brain: Relation to pain pathways. Brain Res 80:135, 1974.
153. Stanley TH, De Lange S: The influence of patient habits on dosage requirements during high dose fentanyl anesthesia. Can Anesth Soc J 31:368, 1985.
154. Shafter A, et al: Use of fentanyl infusion in the intensive care unit: Tolerance to its anesthetic effect. Anesthesiology 59:245, 1983.
155. Stanley TH, Webster LR: Anesthetic requirements and cardiovascular effects of fentanyl-oxygen and fentanyl-diazepam-oxygen anesthesia in man. Anesth Analg 57:411, 1978.
156. Tammisto T, et al: A comparison of the circulatory effects of the analegesics fentanyl, pentazocine and pethidine. Br J Anesth 42:317, 1970.
157. Liu WS, et al: Cardiovascular dynamics after large doses of fentanyl and fentanyl plus N_2O in the dog. Anesth Analg 55:168, 1976.
158. Holmes CM: Supplementation of general anesthesia with narcotic analgesics. Br J Anesth 48:907, 1976.
159. Bailey PL, et al: Small doses of fentanyl potentiate and prolong diazepam induced respiratory depression. Anesth Analg 63:183, 1984.
160. Corssen G, Domino EF, Sweet RB: Neuroleptanalgesia and anesthesia. Anesth Analg 4:748, 1964.
161. Grell FL, et al: Fentanyl in anesthesia: A report of 500 cases. Anesth Analg 49:523, 1970.
162. Hill AB, et al: Prevention of rigidity during fentanyl-oxygen induction of anesthesia. Anesthesiology 55:542, 1981.
163. Pasternak GW, Childers SL: Opiates, opioid peptides and their receptors in critical care: State of the art, in Shoemaker WC (ed): Society of Critical Care Medicine: Textbook of Critical Care. Philadelphia, Saunders, 1984, pp. 1–60.
164. Bailey PL, Stanley TH: Pharmacology of intravenous narcotic anesthetics, in Miller RD (ed): Anesthesia, ed 2. New York, Churchill Livingstone, 1986, pp

745–797.

165. Tanaka GY: Hypertensive reaction to naloxone. JAMA 228:25, 1974.

166. Azar I, et al: Cardiovascular responses following naloxone administration during enflurane anesthesia. Anesth Analg 60:237, 1981.

167. Taff RH: Pulmonary edema following naloxone administration in a patient without heart disease. Anesthesiology 59:576, 1983.

168. Purdell Lewis J: Studies of fentanyl-supplemented anesthesia: Effects of naloxone on circulation and respiration. Can Anesth Soc J 27:323, 1980.

169. Singer SJ, Nicholson GL: The fluid mosaic model of the structure of cell membranes. Science 175:720, 1972.

170. Strichartz G: Molecular mechanism of nerve block by local anesthetics. Anesthesiology 45:421, 1976.

171. De Jong RH, Heauner JE: Diazepam prevents local anesthetic seizures. Anesthesiology 34:523, 1971.

172. Mather LE, Cousins MJ: Local anesthetics and their current clinical use. Drugs 18:185, 1979.

173. Selden R, Sasahara AA: Central nervous system toxicity induced by lidocaine. Report of a case in a patient with liver disease. JAMA 202:908, 1967.

174. Savarese JJ, Covino BG: Basic and clinical pharmacology of local anesthetic drugs, in Miller RD (ed): Anesthesia, ed 2. New York, Churchill Livingstone, 1986, pp 983–1013.

175. Koven IH, et al: Correction by hyaluronidase of interstitial tissue transport defect during shock: A new approach to therapy. J Trauma 15:992, 1975.

176. Grossman JAI: The effect of hyaluronidase and dimethyl sulfoxide (DMSO) on experimental skin flap survival. Ann Plast Surg 11:3, 1983.

177. Glogau R: Sterile injection for the wet technique in liposuction. Presented at Joint Live Workshop on New Procedures and Techniques in Cosmetic Surgery, Department of Cosmetic Surgery, The Graduate Hospital, Philadelphia, July 1985.

178. Jenkins MT, et al: The postoperative patient and his fluid and electrolyte requirements. Br J Anesth 47:143, 1975.

179. Heatin H, et al: Incidence, pathomechanism and prevention of dextran-induced anaphylactoid/anaphylactic reactions in man. BioStandards 48:179, 1980.

180. Doemicke A, et al: Blood and blood-substitutes. Br J Anesth 49:681, 1977.

181. Metildi LA, et al: Crystalloid vs. colloid in fluid resuscitation of patients with severe pulmonary insufficiency. Surg Gynecol Obstet 158:207, 1984.

182. Miller RD, Brzica SM Jr: Blood, blood components, colloids and autotransfusion therapy, in Miller RD (ed): Anesthesia, ed 2. New York, Churchill Livingstone, 1986, pp 1329–1367.

183. Virgilo RW, et al: Crystalloid vs. colloid resuscitation, is one better? Surgery 85:129, 1979.

184. Shires GT, Canizara PC: Fluid electrolyte and nutritional management of the surgical patient, in Schwartz SI (ed): Principles of Surgery, ed 3. New York, McGraw-Hill, 1979, pp 65–97.

185. Giesecke AH Jr, Egbert LD: Perioperative fluid therapy—Crystalloids, in Miller RD (ed): anesthesia. ed 2. New York, Churchill Livingstone, 1986, pp 1313–1328.

186. Mladick RA: Sixteen months experience with the Illouz technique of lipolysis. Ann Plast Surg 16:220, 1986.

187. Newman J, Dolsky RL: Complications and pitfalls of facial lipo-suction surgery. Am J Cosmetic Surg 2:1, 1985.

187a. Fredricks S (chairman), Anastasi GW, Baker JL, et al: Five-year updated evaluation of suction-assisted lipectomy. Report of the Ad-hoc Committee on New Procedures, American Society of Plastic and Reconstructive Surgery, September 30, 1987.

188. Wood EC, Becker PD: Beard's Massage, ed 2. Philadelphia, WB Saunders, 1974.

189. Summer W, Patrick MS: Ultrasonic Therapy. New York, Elsevier, 1964.

190. Lewis EM, Pruitt M: Massage and ultrasound, in Hetter GP (ed): Lipoplasty: The Theory and Practice of Blunt Suction Lipectomy. Boston, Little, Brown,

1984, pp 175–178.

191. Nichter LS, et al: Rapid management of persistent seromas by sclerotherapy. Ann Plast Surg 11:3, 1983.

192. Goldzer RC, et al: Intrapleural tetracycline for spontaneous pneumothorax. JAMA 241:7, 1979.

193. Wallach HW: Intrapleural tetracycline for malignant pleural effusion. Chest 68:510, 1975.

194. Elam MV, Berkowitz F: Submental and submandibular lipectomy by liposuction surgery. Am J Cosm Surg 1:1, 1984.

195. Aronsohn RB: Liposuction of the naso-labial fold—A preliminary report. Am J Cosmetic Surg 1:2, 1984.

196. Plot S: Lipo-aspiration et tunnelisation dans la chirurgie de visage. Rev Chir Esth Langue Fr 10(39):9, 1985.

197. Plot S: Lifting by lipo-aspiration. Presented at the French Society of Esthetic Surgery, Paris, June 1982.

198. Plot S: Nouvelles conceptions du traitement chirurgical dans les liftings cervico-faciaux. Rev Chir Esth Langue Fr 8(31):7, 1983.

199. Plot S: Lipoaspiration et tunnelisation dans la chirurgie du visage. Rev Chir Esth Lague Fr 10(33):9, 1983.

200. Newman J, Fallick H: Liposuction tunneling in conjunction with rhytidectomy. Am J Cosmetic Surg 1:3, 1984.

201. Newman J, et al: Introduction of the facial lipo-spatula extractors. Am J Cosmetic Surg 2:2 1985.

202. Vistnes LM: Anatomical consideration in aesthetic plastic surgery of the anterior neck, in Kaye BL, Gradinger GP (eds): Symposium on Problems and Complications in Aesthetic Plastic Surgery of the Face. St. Louis, Mosby, 1983, pp 269–273.

203. Pitanguy I: Trochanteric lipodystrophy. Plast Reconstr Surg 35:280, 1964.

204. Fournier PF, Asurey NL: Le Pitanguy fermé. Rev Chir Esth Langue Fr 11(42):57, 1986.

205. Elam M: Liposuction contouring of the knee and ankle. Am J Cosmetic Surg 1:2, 1984.

206. Asken S: Liposuction surgery as performed under local anesthesia in an outpatient surgical facility. Presented at the Second International Congress of Aesthetic Surgery, Tokyo, May 1985.

207. Dolsky RL, Asken S, Nguyen A: Surgical removal of lipoma by liposuction surgery. Am J Cosmetic Surg 3:3, 1986.

208. Aronsohn RB: Lipo-suction of the naso-labial fold—A new pressure device. Am J Cosmetic Surg 3:4, 1986.

209. Neuber F: Fettransplantation. Chir Kongr Verhandl Dsch Gesellsch Chir 22:66, 1893 [in German].

210. Neuhof H: The Transplantation of Tissues. New York, Appleton & Co., 1923.

211. Gurney CE: Experimental study of the behavior of free fat transplants. Surgery 3:680, 1938.

212. Peer LA: Transplantation of fat. Presented at the Annual Meeting of the American Association of Plastic Surgeons, Boston, June 1948.

213. Peer LA: Transplantation of fat, in Converse JM, McCarthy J and Littler JW (eds): Reconstructive Plastic Surgery. Philadelphia, WB Saunders 1977 pp 251–261.

214. Stevenson TW: Free fat grafts to the face. Plast Reconstr Surg 4:458, 1949.

215. Laico JE: Fat grafting in facial deformity. Philippine J Surg 6:226, 1951.

216. Noele R: Fat-pad sliding and fat grafting for leveling lid depressions. Clin Plast Surg 8:4, 1981.

217. Ellenbogen R: Free autogenous pearl fat grafts in face—A preliminary report of a rediscovered technique. Ann Plast Surg 16:179, 1986.

218. Bames HO: Augmentation mammoplasty by lipo-transplant. Plast Reconstr Surg 11:404, 1953.

219. Illouz YG: Réutilization de la graisse après liposuction. Presented at the 3rd Meeting of the International Society of Surgical Body Sculpturing and Lipolysis, Paris, June 1984.

220. Illouz, YG: L'avenir de la réutilization de la graisse après liposuccion. Rev Chir Esth Langue Fr 9(36):13, 1984.

221. Illouz YG: L'Avenir de la reutilization de la graisse après liposuction. (Suite) Rev Chir Esth Langue Fr 10(38):19, 1985.

222. Cotton FJ: Contribution to technique of fat grafts. N

Engl J Med 211:1051, 1934.

223. Chajcir A: Utilization of Liposuction fat for body defects. Workshop on New Procedures and Techniques in Cosmetic Surgery, The Graduate Hospital, Department of Cosmetic Surgery, Philadelphia, July 1985.

224. Illouz YG: De l'utilization de la graisse aspirée pour combler les défects cutanés. Rev Chir Esth Langue Fr 10(40):13, 1985.

225. Fournier PF: Microlipoextraction et microlipo-injection. Rev Chir Esth Langue Fr 10(41).

226. Johnson GW: Body contouring by macroinjection of autogenous fat. Presented at the First World Congress of the American Academy of Cosmetic Surgery, New Orleans, October 1986.

227. Fischer G: Autologous Fat Implantation for Breast Augmentation. Presented at the Workshop on Liposuction and Autologous Fat Reimplant, Isola d'Elba, September 1986.

227a. Bircoll MJ: New frontiers of suction lipectomy. Presented at the Second Asian Conference of Plastic Surgery, Pattyia, Thailand, February 1984.

227b. Bircoll, MJ: Cosmetic breast augmentation utilizing autologous fat and liposuction techniques. Plastic and Reconstruction Surgery 79(2).

228. Otteni FM: La lipoplastie d'augmentation par autogreffes d'îlots adipocytares ou "lipo filling." Rev Chir Esth Langue Fr 11(45):27, 1986.

229. Vazquez-Barbe, et al: Histological study of aspirated fat. Am J Cosmetic Surg 2:4, 1985.

230. Saunders MC: Survival of autologous fat grafts in human and mice. Connect Tissue Res 8:85, 1981.

231. Chajchir, A, Benzaquen I: Liposuction fat grafts in face wrinkles and hemifacial atrophy. Aesthetic Plast Surg 10:115, 1986.

232. Tcheupdjian L: The specificity of the adipocytes in the human body. Presented at the Workshop on New Procedures and Techniques in Cosmetic Surgery, The Graduate Hospital, Department of Cosmetic Surgery, Philadelphia, July 1985.

233. Vilain R: Prevention and treatment of waves after suction lipectomy. Ann Plast Surg 17:3, 1986.

234. Asken S: Local anesthesia for microlipoextraction and microlipoinjection. First International Workshop on Liposuction and Autologous Fat Transplantation Under Local Anesthesia, Westport, Conn., August 1986.

235. Moskowicz L: Treatment of facial hemiatrophy by transplantation of fat tissues. Med Clin 26:1472, 1930.

236. Fournier PF: Hasard et lipo extraction; reflections et lipoplastie. Rev Chir Esth Langue Fr 11(42):51, 1986.

237. Newman J: Preliminary report on "fat recycling": Liposuction fat transfer implants for facial defects. Am J Cosmetic Surg 3:2, 1986.

238. Manstein CH, et al: A filter and collection device for suction assisted lipectomy. Ann Plast Surg 16:538, 1986.

239. Asken S: Microlipoextraction and injection for correction of certain skin defects and signs of aging. Presented at the Third Annual Scientific Meeting of the American Academy of Cosmetic Surgery and the American Society of Liposuction Surgery, Los Angeles, February 1987.

240. Asken S: Autologous fat transplantation—Micro and macro techniques. Am J Cosmetic Surg [special issue on fat transplantation ("recycling")] 4(7):111, 1987.

241. Dolsky RL, et al: Adipocyte survival. Presented at the Third Annual Scientific Meeting of the American Academy of Cosmetic Surgery and the American Society of Liposuction Surgery, Los Angeles, February 1987.

242. Campbell G LeM, Newman J: Lipo-suction research findings of the various factors determining fat survival in liposuction implantation. Presented at the Third Annual Scientific Meeting of the American Society of Liposuction Surgery, Los Angeles, February 1987.

243. Viñas J, et al: La region pubiana coma dadora de injertos de grasa. Prenat Med Arg 60:942, 1973.

244. Katoes AS Jr, et al: Perfused fat cells: Effects of lipolytic agents. J Biol Chem 248:5089, 1933.

245. Sidman RL: The direct effect of insulin on organ cultures of brown fat. Anat Rec 124:723, 1956.

246. Olafsky JM, Chang H: Insulin binding to adipocytes—Evidence for functionally distinct receptors. Diabetes 27:940, 1978.
247. Smith U: Human adipose tissue in culture studies—On the metabolic effects of insulin. Diabetologia 12:137, 1976.
248. Soloman SS, Duckworth WC: Effect of antecedent hormone administration on lipolysis in the perfused isolated fat cells. J Lab Clin Med 88:984, 1976.
249. Soloman SS: Comparative studies of antilipolytic effect of insulin and adenosine in the perfused isolated fat cell. Horm Metab Res 12:601, 1980.
250. Menschik Z: Vitamin E and adipose tissue. Edinb Med J 51:486, 1944.

Index

Note: Page numbers in italic indicate material in figures or figure legends or tabular material.